The Glycine Miracle

The Science of Living Long and Free from Inflammation

by

Joel Brind, PhD

The Oaklea Press

WWW.OAKLEAPRESS.COM

Table of Contents

Introduction

Americans spend more money than ever on healthcare, so why does the average American seem less and less healthy? Average American life expectancy has actually started to go down in the last couple of years.

Putting aside the epidemic of obesity, even in childhood, the average American seems pretty healthy until about the age of 40. But then the progression begins of medications and surgeries that mark the decline in health from age 40 on. Everywhere Americans are taking meds for prediabetes and diabetes, high blood pressure, arthritis, psoriasis and cancer, among many others. Many need to undergo drastic interventions like knee and hip replacements by the time they are 60, as well as cancer surgeries and vascular surgeries like stents, bypasses and heart valve replacements.

So the average American manages to hobble up to about age 75 or 80 before he or she succumbs, usually to some chronic illness like heart disease or cancer. Only a minority of Americans actually can be said to die of old age, even among those who may reach their nineties.

What's wrong with this picture? Why don't most Americans live to be 100 or more, free of chronic illness to the end?

Many healthcare professionals, especially alternative health advocates, believe the answer lies in correcting nutritional imbalances, deficiencies or toxicities, and they are right.

Most medical and medical research experts are now coming around to the view that most chronic illnesses, including cardiovascular disease, Alzheimer's Disease and Cancer, are the result of chronic inflammation. They are also right.

The experts also believe that the problem is very complex, due to a complex interplay of genetic, environmental and dietary factors. Here, they are wrong.

Many years ago I learned that when an expert claims that the problem is complex, he really means that he does not know. Complexity then, is often a euphemism for ignorance, but one which preserves the outward appearance of wisdom and expertise. Great discoveries are made when a simple yet fundamental truth is revealed, in the face of which the experts dissolve away. Thus the many experts in different modalities of treating the dreaded scourge of syphilis less than a century ago, disappeared with the discovery of penicillin, just as many complex theories of physics disappeared with the elegant simplicity of $e = mc^2$.

This book is written to share the discovery of the great and simple truth that has been revealed to me through my scientific research, which has been my profession for more than half a century. This truth solves the mystery of chronic illness and early demise that currently plagues our society.

Chapter One
Self-Experimentation Surprises

It was a picture perfect day for baseball. It was back in 2010 in the new Yankee Stadium in New York, and a great day to be wearing my shorts and my NY Yankees tee shirt as the morning chill gave way to the warming mid-June sun in the cloudless blue sky. It was my first time in the new stadium, there with some of my family, as my brother-in-law had scored some really good box seats a few rows behind the first base dugout. As a life-long Yankee fan—as my father had been—I was thrilled to watch the game. And once the game started, I was totally absorbed—especially because the Yankees were winning, and the view was wide and totally unobstructed.

Of course, the sun was also totally unobstructed, and sometime during the fifth inning—about an hour and a half into the game—I noticed a warm sensation on my thighs. I looked down and came to the sudden realization that I was red as the proverbial boiled lobster, everywhere that my fair skin was not clothed. From the hem of my shorts down to my socks, my whole arms and from the neck up, I was sunburned as I had not been since at least my twenties. You see, the thought of putting on sunscreen or sunblock had never even occurred to me in preparation for this day. After all, I wasn't going to the beach!

I immediately segued to a seat in a shaded area some rows back, but I knew, I mean I ***knew***, with 100% certainty, that for the next couple of days, even the simple tasks of dressing, undressing and bathing would be excruciating. And after all that, the dead skin would just blister and peel off, without even leaving the benefit of any significant tanning. I knew this because I had done this to myself several times during my youth, that is, I had gone to the beach and been exposed to the sun for an hour or two or more, with inadequate sunscreen.

This time would be no different, I was certain. And I forgot all about the game. I'm pretty sure the Yankees won, but I don't remember anything else about the game, or even much about the interaction with my family there that day. In fact, there was nothing really memorable that happened. But what ***didn't*** happen was truly a game changer.

You see, during that time I was in the process of conducting an experiment on myself. Starting in 2008 I began taking, as a dietary supplement, 10 grams per day of the amino acid glycine. Exactly why I decided upon this experimental dietary regimen will be explained a bit later in this book. The fact is, I did not know quite what to expect.

I really didn't expect anything dramatic to happen. Rather, I had come to the idea that taking that much glycine every day should be of some long-term health benefit, and that it certainly would do me no harm. After all, glycine, the simplest of all the amino acids, is a bulk nutrient of which everyone takes in a quantity of some number of grams per day. Glycine is also such a small, water-soluble molecule that it cycles through the body in a matter of hours; it does not build up in the long term to any sort of toxic level. I also knew that the typical diet contains about two to three grams per day of glycine, so that, in order to make a significant difference, only a big amount like an additional ten grams per day could be expected to have any noticeable effect.

So, in order to embark on this experimental regimen, I first needed to figure out how to formulate an amino acid powder in such a way that I could easily take 10 grams a day. Glycine is available in capsule form from a number of health food suppliers online. But I would need to take sixteen 500 mg capsules, or eight one-gram capsules every day. That seemed to me too much like a chore. But I had noticed early on, when trying things out informally, that glycine has a sweet taste, albeit with an unpleasant aftertaste. (The name glycine is derived from the Greek, meaning "sweet amine"). I found that the taste was considerably improved when cut with sugar, or added to something like fruit juice which contains natural sugar. Since gelatin also has a high content of glycine in protein form, I settled upon a "glycine Jell-o"-type formulation, wherein each (four-ounce) serving would provide about half a gram just from the gelatin. Then, to unflavored gelatin, I would add natural tart cherry juice. Of course, the cherry juice has its own natural sugar, but since it is naturally tart one could add quite a bit of sweetness without making it too sweet. So my daily regimen ended up being a home-made gelatin dessert, just following the directions on the unsweetened gelatin box with tart cherry juice, only adding 9.5 grams per serving of pure glycine powder. It was a good-tasting dessert and was thus no problem eating it once a day, day after day after day, religiously.

Now, back to the ball game and the sunburn. As I quickly moved to a shady seat to watch the last few innings of the game—having made suitable apologies to my brown-skinned family for the need to do this—I kept beating myself up over how I could have been so stupid as to let this horrible sunburn happen. But as I settled back to watching the game, I soon forgot about the sunburn. In fact, by the time the game was over and we were leaving the ballpark, the redness had already diminished noticeably. And in the succeeding hours, it just continued to fade, and was completely gone—except for just a little bit of tan—by the next morning.

By this time I had read enough about glycine to know that it has some benefit in terms of inflammation. This knowledge was largely amassed by a prolific research team at the University of North Carolina (UNC) headed by a toxicologist named Ron Thurman. Dr. Thurman had unfortunately passed away in 2001 at the age of only 59, from a massive heart attack. Although there were several papers that were published during the ensuing several years, the work at UNC did not continue—as far as I know—and other members of the team were scattered elsewhere.

The Thurman group laid the groundwork for understanding glycine as the immune system's own natural regulator of inflammation. Until then, glycine—acting as the free amino acid, rather than as a building block of protein molecules—was understood to be an inhibitory neurotransmitter in the central nervous system (CNS, i.e., brain and spinal cord). As an inhibitory neurotransmitter, it prevents hyperactivation in neural pathways. That's why glycine supplements have traditionally been sold as sleep aids. It doesn't make you sleepy, but it can help allow you to get to sleep more easily. What the Thurman group found was that the very same molecule on the cell surface membrane of the CNS neurons (nerve cells)—the glycine receptor—was also to be found on the surface of macrophages (the immune cells that actually generate inflammation). They found that the activation of macrophages to produce inflammation was inhibited by adding adequate concentrations of glycine to the cellular environment. The work of the Thurman group was published in a review published in 1999. More on their work later in this book.

What's important for now is the fact that—as I realized by doing some reading in the medical literature to refresh my knowledge on skin physiology, and having done research on the skin during the 1970s and 80s—the painful and damaging aspects of sunburn are actually the result of inflammation; the immune system's reaction to the sunburn, rather than the sunburn itself. When you think about it, it really does make sense. After all, if the sun just makes your skin turn red, why ***doesn't*** it just go away when the offending stimulus—intense sunlight—is withdrawn? What is the point of all that pain and damage that happens in the succeeding hours and days?

So what did ***not*** happen to me this time around was, to say the least, a big surprise. In fact it was one of two big surprise findings in my grand self-experiment with glycine. A bit later that summer, in early July, we were scrambling to finish some work in our house to make it suitable to put up several family members for a couple of days, as our daughter was set to get married at the end of the month.

In particular, I needed to finish a bit of drywall work in the basement. It only required two or three sheets' worth of drywall, but drywall comes in 4-foot by 8-foot sheets, too big to fit into our SUV. The best way around that problem was to make some measurements such that the drywall could be cut in the store into smaller pieces. So my wife and I went to our local Home Depot with a tape measure, a yardstick, a pencil and utility knife. The drywall was stacked up on the concrete floor to a height of about 4 feet. I would climb onto the top of the stack, measure and mark the top sheet for the cuts, score the cuts and break the cut pieces, finish the cuts with the knife and pass down the cut pieces of drywall to my wife. Then I would jump down to the floor and the two of us would load each piece onto the cart. This process would be repeated until we had all the pieces we needed to be cut, cut.

Except I did something wrong. I must have gotten momentarily distracted and turned away from the sheet of drywall I was cutting, and accidentally stepped right off the stack. I fell the full four feet down onto the concrete floor directly onto my tailbone! My wife looked on in horror, speechless, but already imagining the trip to the ER that would surely ensue. And it hurt like hell! But only for about half a minute, after which the pain subsided enough for us to finish what we were doing, load the drywall into the SUV and drive home.

Trouble is, we were under a bit of time pressure that afternoon, as we had bought tickets to a dinner dance that evening. So we went to the dinner-dance anyway, and surprisingly, my injuries were not bothering me that much. So I thought we could try a dance or two, and in fact, it was no trouble at all. More surprisingly, I experienced no pain whatsoever the next morning. Fortunately, I had evidence of the very serious fall I had suffered, in that I did develop a massive bruise on my lower back, and that bruise took a week or so to go away.

So there were no two ways about it: Between the sunburn and the fall at Home Depot, I had experienced two events that should have produced massive inflammation, with all the accompanying pain and disability. But they didn't. And I don't think in my entire career as a scientist that I have ever had any experiment produce such dramatic and definitive results: Inflammation, always known to everyone to be a normal consequence of injury, did not happen after these very substantial injuries.

Now the big question, what did these dramatic results mean?

Chapter Two
Amino Acid Relationships

My graduate degree is a PhD in Basic Medical Science, with specialization in biochemistry, physiology and immunology. Inflammation is a function of the immune system, and thus within the purview of immunology, but my entry into the study of glycine actually began in the arena of nutritional biochemistry and aging.

During the early years of this century, in addition to my full-time professorship at Baruch College (City University of New York), I was working part-time as a biochemist consultant for the Orentreich Foundation for the Advancement of Science (OFAS), a private non-profit research organization for which I had worked full time back in the early 1980s. Other researchers at OFAS had been doing research on the essential amino acid methionine since the early 1990s. In 1993 they published a paper describing the significant extension of normal lifespan of laboratory rats, merely by restriction of the methionine content of the diet they were fed. Since methionine is an essential amino acid, the animals would not survive a diet completely devoid of methionine, but if you gave them just enough to sustain life—while feeding them *ad libitum* (i.e., to eat all they wanted)—they would live 30-40% longer than rats fed the normal control diet. The restriction needed to be severe—about 80% less than normal methionine content—such that, although the rats lived substantially longer, they did not grow to normal adult size. Hence, they stayed pretty much the same size they were when weaned, which was when they were started on the methionine restricted (MR) diet.

The original experiment showing lifespan extension by methionine restriction (MR) turned out not to be a one-off, but the first of many papers on rats and mice by the OFAS group and others in the field. By the early years of this century, MR had been established as a real phenomenon. But the basis of MR remained a mystery. Part of my job was to solve the mystery, and that was my start on the road to glycine.

At this point it would be useful to expand a bit on amino acids in general, to provide some perspective on the connection between these two: glycine and methionine. Most people are familiar enough with the fact that amino acids are the building blocks of proteins. They are small molecules that fit together like Lego blocks, with 20 different ones altogether making up the full variety of all proteins in the body. With the variety provided by the 20, the diversity of protein structures and functions that can be made is almost limitless. There is the wide array of enzymes—proteins that function as "nano" machines that do all manner of chemical work, structural proteins such as collagen—the actual "threefold cord" with which the body is knit together, transport proteins that carry various fats, vitamins and minerals through the bloodstream to the various cells and tissues of the body, hormones that carry messages from one part of the body to another, and a host of others. Twelve of these amino acids are considered "non-essential," since the body (mainly the liver) can make them from simpler compounds, and eight of the amino acids are considered essential, because the body cannot make them from simpler compounds, and thus needs to obtain them from the diet.

There is in the field of nutrition, however, a tendency to oversimplify the amino acids, as if their role in constituting proteins were their only function. In classifying them as essential v. non-essential, their importance in the diet is also oversimplified. Thus there is the pervasive concept of "high quality protein," a term applied to a protein source which is particularly rich in the essential amino acids. Not surprisingly, these have gotten most of the attention in terms of dietary research over the years. Hypotheses concerning the origins of many disease states have sought to find deficiencies in essential amino acids—such as methionine—being responsible for such conditions as cancer. After all, why would anyone waste time and resources studying nutrients that are non-essential?

But all amino acids are not alike, and the "essential" v "non-essential" designation is an oversimplification which can be misleading. Methionine restriction being a good thing is one glaring example of this. Amino acids, you see, serve not only as the building blocks of the proteins. They also serve many other functions, such as being intermediate compounds in the transformation of one amino acid into another, and also building blocks of other, non-protein substances such as DNA. And some amino acids—even just among the essential ones—are more critical than others. Thus it has long been known that methionine has a critical function as the universal methylator; a donor of one-carbon units in the synthesis and modification of many different substances such as DNA, neurotransmitters and other types of compounds. Because of its importance and its essential nature, the body has a number of redundant mechanisms for the recycling, salvage and regeneration of methionine. The methionine cycle, for example, is well known to all students of biochemistry. But precisely because the body is so frugal in its use of methionine, it actually needs very little of it: about 300-500 mg per day; more like the daily requirement for a vitamin rather than a bulk nutrient like an amino acid. And the distribution of methionine in animal proteins is quite asymmetric, with muscle being very methionine-rich, and the collagen of the bones and connective tissues being very methionine-poor. Hence, muscle meats are universally considered "high quality protein," whereas collagen (gelatin) is considered a very low quality protein.

Now take a look at the modern, Western diet, with the typical bacon-and-eggs breakfast, tuna fish or meat loaf sandwich lunch and steak or chicken dinner, seven days a week, and you note that it is extremely methionine-rich. Unfortunately, excess methionine is not harmless, and the body gets rid of it as a toxin when there is too much. Research over the last 20 years has demonstrated that when one consumes a typical high methionine meal, the liver actually does the opposite of its methionine conservation routine, and gets rid of the excess methionine quite rapidly. But there's a catch. While the liver has many pathways available to conserve and recycle methionine, it has only one pathway to get rid of the excess. This makes sense when one understands that the body is designed to better withstand famine than feast. And it turns out that the only pathway for methionine clearance uses up two molecules of glycine for every one molecule of methionine cleared. That would not be a problem of course, were glycine actually non-essential. But that turns out not to be the case. (Much of the biochemistry was worked out in detail in the early years of this century in the laboratory of the late eminent biochemist Conrad Wagner at Vanderbilt University in Tennessee.)

You see, in addition to all the various biochemical functions of glycine, including its role as a building block for proteins, it also has a function that is not really biochemical, meaning that it does not participate in this role, in any biochemical reactions. Specifically, glycine acts as a regulator of cellular function by stabilizing the cell membrane of several types of cells—most importantly, all the various types of macrophages, the cells of the immune system that generate inflammation. Importantly, for this regulatory function, the concentration of glycine in the blood plasma and the fluid that bathes all the cells of the body, needs to be at least three or four times higher than the concentration needed to make all the proteins that glycine is a part of.

As a consequence of eating the typical high animal protein meal (rich in muscle meats and poor in bone and connective tissue, that is), the body actually loses glycine. Hence, the blood plasma of meat-eaters contains ***less*** glycine than that of vegans or vegetarians, even though the total consumption of glycine (and all the other amino acids) is higher. (It is an interesting commentary on the overall state of medical science to note that when this difference was confirmed by Schmidt, et al., a group at Oxford University, and published in 2016, the authors found it counterintuitive, and could not explain it! More on this in Chapter Six).

Getting back to the question of exactly why methionine restriction should extend life, my thinking took a different tack than the one generally followed. Specifically, I thought, "Why do we think of an experimental diet resulting in longer-lived animals to be an extension of lifespan at all?" Rather, why not consider that the shorter-lived animals have their natural lifespan cut short by whatever it is—or isn't—in the control diet that makes them die sooner than those animals on the experimental (in this case, methionine-restricted) diet?

Chapter Three
The Cells that Do Inflammation

What exactly is the nature and purpose of inflammation? Inflammation may well be the most familiar function of the immune system, inevitably popping up with any sort of injury, for example. Its classic description goes back centuries in the medical textbooks, i.e., "rubor, et tumor et calor et dolor" (Latin for redness and swelling and heat and pain). When occurring at any joint, immobility is added to the list of symptoms. And of course everybody knows that the first aid treatment for blunt injury is to put an ice pack on the injured area, for the express purpose of suppressing inflammation. This bit of common wisdom raises an interesting question, namely, why does the body normally respond to injury in an inappropriate way, such that this natural response—inflammation—needs to be suppressed? Perhaps this response is not so natural after all, rather suggesting some underlying disorder?

Now that medical science has elucidated bodily functions at the cellular and biochemical level, we can take a closer look at what actually happens in inflammation. In terms of the immune system, inflammation is seen to be a function of what is called **innate immunity**, i.e., a relatively non-specific response to infection or cellular damage. The body needs to respond to infection quickly and vigorously, lest the invading microbes spread and multiply and kill the host before a specific antibody response—which takes weeks to muster—can be mobilized. Biochemically, that response includes the manufacture and deployment by the immune system of a host of toxic chemicals (collectively known as cytokines) which can kill the invading microbes. But since these chemicals are non-specific poisons, they also do damage to normal bodily tissues. This manifests in part, as the above-described familiar symptoms of inflammation. This response thus raises another interesting question, namely, what is the point of secreting poisons to kill invading microbes in circumstances where there are no microbes to kill, such as in blunt injury? One common hypothesis suggests that the immobility of injured joints protects the joint from further injury; sort of a natural cast or splint. Indeed it is generally thought that inflammation is part of, or essential for the initiation of the healing process. The trouble is, all those cytokine poisons actually inhibit healing. So there is really no satisfactory explanation out there for why the body essentially engages in a self-destructive process in response to injury. Indeed, the big hunt is on for an explanation as to why inflammation seems to happen chronically, often seemingly for no reason at all. A recent review on the subject of chronic inflammation in the popular journal *Epoch Times* describes the situation thus: "When the body is infected or injured, inflammation—often likened to fire—is nature's way of burning away pathogens and repairing damage. Once the threat is eliminated, inflammation should subside. But if the fire continues to smolder, it can become a chronic issue." Here is embodied the central contradiction: How can we speak of inflammation as a proper response to injury, but one which needs to subside "once the threat is eliminated," if there is no threat to begin with? So it is clear that the wrong question—i.e., "Why does inflammation persist after the initial response?"—is being asked. To answer it, a distinction is made between "normal inflammation" and "chronic inflammation," the latter of which is inflammation that persists more than three months. But this distinction is arbitrary and artificial. As we shall see later in this book, short-term inflammation, when inappropriate, can be just as damaging—even fatal—as chronic inflammation.

But at least it is now widely understood that inflammation is the common denominator for most of the chronic diseases that make people sick and die these days, from arthritis to diabetes to cardiovascular disease to cancer.

So let's take a look at the cellular players in the inflammatory process. The cells that actually generate inflammation are the immune system's first responder cells, collectively called macrophages (derived from the Greek, meaning "big eaters"). These cells all originate in the bone marrow, where all of the red and white blood cells originate. The name macrophage refers to these cells acting like amebas, eating up dead and dying cells and cellular debris and invading microbes by the process of phagocytosis. Phagocytosis is the process in which the cell doing the eating actually moves its membrane and cytoplasm around to engulf large particles. Once engulfed, the contents of the newly formed food vacuole are digested by cellular enzymes and chemically recycled, if they are digestible. If these particles are not digestible, they simply remain inside the macrophages permanently. This is the case, for example, of the pigment particles used in tattoos. The fact that the macrophages remain alive and therefore capable of some movement, is why tattoo images gradually become blurred with time.

Several types of macrophages circulate as white cells in the blood (i.e., some of the monocytes, but for simplicity, I also include the more abundant white blood cells called neutrophils, as well as basophils and eosinophils in the macrophage category), and there are also macrophages which are resident in perhaps all the organs and tissues of the body, at least wherever they have been looked for. Hence, there are known to be macrophages specific to the liver and the lungs and the bone and the skin and the brain, for example. And even though these macrophages typically look like the cells of the organs in which they are resident—save for chemical surface markers by which they can be distinguished chemically—they all originate in the bone marrow, migrating to the other organs during the embryonic stages of life.

Rend or Mend?

The macrophages thus comprise the immune system's first responder cells, and they get activated whenever the chemical signature of a disease-causing microbe is detected or anytime there is any tissue damage (cellular death). But their activation is often misunderstood. Recent research has demonstrated that macrophages can be activated in two ways: one level of activation causes them to migrate to the site of injury and phagocytize the dead cells and cellular debris. Macrophages which are activated to engage in this activity are typically referred to as "M2." But it takes a different, second type of activation, typically referred to as "M1" (sometimes called "priming") which causes them to initiate the process of cytokine production and secretion characteristic of inflammation. To reflect this dichotomous type of activation, we may refer to M1 as "rend" mode, and M2 as "mend" mode. So is a macrophage involved in dealing with tissue injury there to rend or to mend?

A good metaphor for the action of macrophages is the action of first responders in society. For example, when there is an accident on a freeway or expressway, the first responders that show up are the police. They arrive at the scene of the accident, call in backup and ambulances and tow trucks, take accident reports from witnesses and redirect traffic to keep it flowing. When their work is complete, they return to their nearby bases and traffic flow returns to normal. These would all be considered "mending" actions. Of course, being police, all the while they are armed with deadly force—capable of rending, rather than mending—but would not think to draw their weapons and start shooting unless bad actors are present. But just imagine if every time there was an accident on the highway, the police showed up and started shooting? That would be insanely destructive! Yet that is exactly what happens within our own bodies even when there are no bad actors present, i.e., injury—and therefore inflammation—in the absence of infection. No wonder people's physical health begins to go downhill by the time half a normal lifespan has been lived!

But there is much more to the story of what actually happens in inflammation, some of it only elucidated in the last couple of years. As noted earlier, inflammation is basically a nonspecific reaction to the detection of the chemical signature of any type of dangerous microbe. But it is a multifaceted and graded reaction, involving several different types of weaponry. It is not necessary here, however, to get too much into the weeds, so to speak, of all the diverse different types of cytokines; chemical effectors of damage to microbes, some of which are not fully understood. Suffice it to say that macrophages engage in the equivalent of shooting a gun, but it may also escalate to the equivalent of artillery and ultimately, the equivalent of a nuclear weapon!

The simplest action of an M1 activated macrophage is the secretion of molecular "bullets:" small, destructive molecules such as hydrogen peroxide. If you look at a bottle of hydrogen peroxide you can purchase at a pharmacy, it is to be used as a topical antiseptic; not to be taken internally. But internally, M1 macrophages actually secrete hydrogen peroxide for the same purpose: killing microbes that have gotten into the body. These small molecules are secreted by being produced in the macrophage in small vesicles; little bubble-packets that migrate to the cell membrane, with the contents being released into the extracellular space when the membrane of the vesicle fuses with the cell membrane, both membranes having the structure of highly specialized "soap bubbles." There are other, larger effectors of inflammation, cytokine proteins such as Tumor Necrosis Factor alpha (TNFα). TNFα is produced in the cell as a protein molecule that also ends up in a vesicle that fuses with the cell membrane, and it can activate other immune cells by contact with such cells, and it can also have its active part clipped off so that the resulting soluble TNFα is secreted into the fluid outside whence it can migrate to activate other cells elsewhere in the body.

And then, for a greater inflammatory response, there is the construction of an 'artillery piece' called an inflammasome within the macrophage. An enzyme called caspase within the cell initiates the construction of a very large, multi-protein complex in the shape of a cylinder. Activated by another enzyme in the cell called a gasdermin (gasdermin D being the most well studied type), this inflammasome creates a large cylindrical pore in the cell membrane, a literal artillery barrel large enough (10-20 nanometers in diameter) to enable the bulk flow outward of fairly large inflammation effectors such as Interleukin 1beta (IL-1β). IL-1β, for example, has long been known as a pyrogen; a substance which can induce a body-wide fever, for the purpose of killing infecting bacteria.

As you might guess, a number of such large pores in the cell membrane can be pretty destabilizing to the integrity of the macrophage. Such large openings in the cell surface formed by gasdermin D allow for the bulk passage of water molecules, for example. In the bloodstream, the presence of high concentrations of albumin and other proteins keeps these cells from swelling and bursting. But within the tissues outside the bloodstream, where the cells are bathed in essentially protein-free lymph, water flows in, ultimately leading to the cells bursting (which we may liken to the detonation of a nuclear weapon), a process that is termed pyroptosis, and releasing a huge torrent of cellular contents (often referred to collectively as "DAMPS" for "damage-associated molecular patterns," which tremendously exacerbate the inflammatory reaction.

This sort of reaction has long been known to occur in severe infectious reactions. It was first observed in 1986 by Arthur Friedlander at the US Army Medical Research Institute in Maryland, who demonstrated the ability of anthrax lethal toxin to directly destroy macrophages. In 1992 the process was observed in detail by A Zychlinsky of the Pasteur Institute in Paris, Fance, as the induction of suicide in human intestinal macrophages by Shigella flexneri bacteria. First presumed to be an example of a well known process of programmed cell death, called apoptosis, the different mechanism by which this cellular suicide occurs was recognized in 2000 and termed pyroptosis by Brad Cookson and Molly Brennan of the University of Washington in Seattle, who worked with Salmonella typhimurium bacteria. Pyroptosis is maximal and widespread in septic shock, a condition in which an infection has spread to the entire bloodstream, and by provoking the rupture of many macrophages and the release of so many DAMPs, is usually rapidly fatal.

Now this begs the same question we entertained earlier, namely, why does the immune system generate a reaction so destructive that it can kill the whole body itself? Actually, such a violent reaction is appropriate for virulent bacterial infections like Salmonella and anthrax, but septic shock does not usually result because the process is somehow self-limiting. How? Surprisingly, the answer is exactly the same answer that applied to the initiation of inflammation in the case of injury without infection (sterile injury). Again, it is a matter of the concentration of the amino acid glycine!

Just as the Thurman group at UNC had elucidated the role of glycine in the initiation of inflammation via the glycine-gated chloride channel called the glycine receptor, another research team had been investigating, since the 1980s, the protective role of glycine in septic shock. Thus, nephrologist Joel Weinberg and his group at the University of Michigan determined that glycine's protective effect in septic shock does not involve the glycine receptor. But the molecular details of this effect has only become clear from research in the last couple of years, at several major research centers, including that of the pharmaceutical company Genentech in San Francisco, CA.

It turns out that, contrary to the prevailing dogma for decades, the swelling and bursting of the macrophages is not a passive process owing to the pressure of inflowing water. Swelling, yes, but this does not lead to the bursting of cells without the action of another surface protein called "ninjurin 1" (NINJ1). When Nobuhiko Kayagaki and colleagues at Genentech tested laboratory mice from which the NINJ1 genes had been removed (called NINJ1 knockouts), the final step in pyroptosis did not occur; the nuclear weapon did not detonate. In order for the final step of membrane rupture to occur, the NINJ1 protein molecules which normally float within the cell membrane must aggregate together, and glycine prevents that from happening.

This is really quite an astonishing discovery: The lowly "nonessential" amino acid glycine is the very molecule which specifically prevents both the very first step (via the classical glycine receptor) and the very last step (cell membrane rupture, or CMR) of the inflammatory response, so that inflammation is not excessive or inappropriate, unless of course, glycine is deficient. Hence, glycine is both the alpha and the omega of the control of inflammation!

But a related question then arises, namely, how is pyroptosis—which can happen appropriately to combat infection from such pathogenic bacteria as Salmonella and anthrax—ever limited in a body that is glycine-deficient, such that it does not invariably escalate to a life-threatening scenario like septic shock? After all, pyroptosis causes the release of inflammatory mediators which dramatically accelerate damage, so if glycine is low, what stops the situation from spiraling out of control? You may find this surprising; even counterintuitive, but the answer is still glycine! How, exactly, you might ask?

It turns out that all living cells normally concentrate glycine from the surrounding fluid. That is, cells expend energy to pump glycine in, such that the normal concentration of glycine inside of cells is five to ten times higher inside than outside. Hence, when a macrophage "goes nuclear," i.e., when the cell membrane ruptures and releases inflammatory mediators (DAMPS) all over the place, it also substantially increases the local extracellular concentration of glycine. The increased concentration of glycine therefore acts to stabilize other macrophages, both preventing the final stages of pyroptosis, and stabilizing the membranes of unactivated macrophages via the glycine receptor. Finally, the system fails and results in septic shock when glycine concentration is so low, and/or the infection has become so massive, that inflammation passes the point of no return.

Later, we will revisit the work of Joel Weinberg on the subject of how glycine prevents cellular necrosis, a process that is not caused by inflammation, but other insults such as oxygen deprivation. But as we will see, it is virtually the same process of cellular death by necrosis in the kidney cells that Weinberg, et al., study, which happens in pyroptosis in macrophages. In other words, the final detonation and self-destruction that characterizes the ultimate step in inflammation, is really the same process in the death of kidney cells by necrosis. Importantly, the sudden release of cellular contents in necrosis or pyroptosis causes inflammation to be generated in the vicinity of the event. In fact, this is the kind of cellular demise that happens in septic shock, which causes multiple organ failure from massive cell death. The Thurman group demonstrated the ability of glycine to prevent septic shock in animals back in the 1990s, and the Weinberg group demonstrated the protective effect of glycine in cellular necrosis back in 1987!

All of this action I have summarized here has been the subject of astonishing amounts of research in the last few years. For example, merely using the search term "pyroptosis" in review papers—papers which review the findings of many primary research studies—generates 531 hits—531 separate, unique review papers—in the National Library of Medicine (Medline) database! And as we have seen, the knowledge base relating to inflammation and glycine has been expanding rapidly. Hence one might think the realization of the central importance of glycine and glycine deficiency in human health and disease would be imminent. I, for one, certainly hope that it is, and the very purpose of this book is to hasten that day.

Indeed, we may now distill recent discoveries about cellular death and macrophage function to essentially two mechanisms:

1) Apoptosis, which is the orderly condensation of cellular contents into "bite-sized" pieces of molecular aggregates, and the emission of signals to attract macrophages and to activate them to the "mend" or M2 mode. The M2 macrophages then quietly devour the remains of the cell, clearing the way for the normal functioning of the tissue.

2) Pyroptosis (or a similar type of cell death, such as necrosis or ferroptosis), which, as described earlier in this chapter, is the swelling and explosive bursting of the cell, spewing its contents willy-nilly in the vicinity, and attracting and activating macrophages to the "rend" or M1 mode, propagating and magnifying the inflammatory state. This is what results in chronic inflammation, with its resulting pain and tissue damage and even malignant tumor generation, or blood vessel blockage, depending upon where in the body it occurs.

When glycine concentrations are adequate, pyroptosis only occurs in a limited way in response to a virulent infection. But when glycine concentrations are too low, pyroptosis occurs inappropriately in response to sterile injury, and excessively even in response to infection, leading to chronic inflammation and some form of early demise of the whole body.

Unfortunately, as we have also seen in earlier chapters of this book, the focus of researchers is typically pharmacological rather than physiological. In other words, rather than studying a disease process in order to understand normal function and to learn how normal function may be restored, disease processes are studied in order to develop an artificial method—a drug—to correct the diseased condition. A particularly instructive example of this sort of thinking, from the Cookson et al. group at the University of Washington, is embodied in the literal bottom line of a study by Wendy Loomis et al. (including Brad Cookson) on the action of glycine to prevent the final stage of pyroptosis:

"Glycine administration is highly protective in models of sepsis, suggesting that understanding the mechanism of glycine action may provide novel therapeutic targets for inflammasome-mediated pathology. This study identified novel pyroptotic cytoprotectants of much higher potency than glycine, which may be useful for future studies of pyroptosis."

That makes me ask the following question: If glycine, a normal and natural component of the blood plasma and lymph, is the natural regulator which arrests the process of septic shock, why look for something of greater potency? This is not to belittle the importance of the study in helping to characterize and understand the nature of glycine's action in this process, but the focus seems rather, as we have seen before, to come up with new drugs. And after all, as discussed a bit earlier here, the final piece of the puzzle of the glycine mechanism was discovered in the laboratories of a pharmaceutical company (Genentech).

Chapter Four
The Inflammation Switch

A great deal of medical research attention is focused on the question: "What is responsible for inflammation becoming chronic and thus doing damage and causing chronic disease?"

Unfortunately, this bypasses our earlier question about why the body resorts to inflammation in the first place in situations—e.g., blunt injury—when it cannot possibly do any good, and only causes harm. In other words, inflammation is taken as a given in such circumstances, and the dismissal of the real question precludes its ever being answered! (Later in this book, the scientific process itself will be discussed, and the importance of asking the right questions will loom large.) For me, the surprise of not experiencing inflammation when it had been expected as a certainty, forced me to question my basic assumptions.

To examine this question about the switch that turns inflammation on and/or keeps it on inappropriately, we return to the seminal research of Dr. Thurman's group at UNC in the late 1990's. It was the Thurman group that elucidated the likely mechanism by which glycine acts to prevent the inappropriate or excessive priming of macrophages, rather like a trigger lock in our first responders metaphor. To understand this process more specifically, I need to introduce some basic knowledge of cells and cellular physiology.

The cell is the microscopic functional unit of a living organism. Indeed, some organisms—like amebas—consist of just a single cell. In fact, every living body—including the human body—starts out life on earth as a single cell at the time of fertilization or conception. Then it grows by cellular division and specialization. A fully grown human body is composed of some 100 trillion cells of some hundreds of different types, ranging from the epithelial cells of the skin to the various types of blood cells, muscle cells, nerve cells and so on, including all the various types of macrophages of the immune system. The boundary of every cell is formed by what is called the cell membrane or plasma membrane, which differentiates the inside from the outside of the cell. So for example, the blood plasma constitutes the outside environment for all the blood cells. The membrane itself is very fluid and constantly changing, to enable the cell to function in and adapt to its surroundings, and to respond to messages from other parts of the body and other cells nearby. Its structure consists of a soapy film, with highly diverse and specialized molecules floating around in it. Thus, it is rather like a highly complex and specialized soap bubble. Each cell contains within it a nucleus, which contains the DNA; the library of information that specifies everything the cell is capable of making in terms of proteins. Various proteins comprise the cellular machinery itself and everything the cell can make for export for use outside the cell, such as hormones and cytokines. "Organelles" within the cell, called mitochondria, function as power plants to generate the chemical energy the cell requires. Other organelles include ribosomes—which are basically protein knitting machines, endoplasmic reticulum—which are assembly lines composed of many ribosomes, and proteasomes—which are basically protein recycling centers. There are also other organelles that perform various functions. The proteins and other molecules that float around within the cell membrane itself, determine the exchange of materials and messages between the inside and the outside of the cell. Every living cell must maintain what is called "homeostasis," i.e., a constant internal environment, while constantly exchanging energy, material and information with the cells' external environment. Cells such as macrophages may be quiescent—just minding their own business—or they may be activated to perform a particular bodily function, such as inflammation. When activated, a cell changes its internal environment in response to its activation. A macrophage, for example, when it detects—via receptor molecules floating in its cell membrane—the presence of chemicals coming from other cells which have been damaged or killed, goes into phagocytosis (mend or M2) mode. In this mode, it cleans up the mess of dead cells and cell debris, so that healing—via new tissue growth—can occur. But the macrophage can also become primed to enter the rend (M1) mode, when it detects—again, via specific receptor molecules in its cell membrane—the presence of components of bacteria or fungi or some other infectious microbe. That sets into motion the construction of an inflammasome within the cell; the machinery for the production and secretion of various poisons, so that the invading microbes can be killed before they endanger the life of the host.

Interestingly, the very switch that turns on or primes the macrophage is actually an electrical switch, essentially just like a wall switch that turns on the light in a room! When the light switch is off, one pole of the switch carries 120 volts of electricity, whereas the other pole carries zero volts. When the switch is turned on, the two poles are connected by a metal (usually copper) that conducts electricity, thus bringing the voltage difference between the two poles to zero. The electricity reaches the light bulb, and it lights up.

In the case of a living cell, the voltage is created by the fact that the internal environment of the cell is different from the external environment, due to the presence of different concentrations of dissolved ions. Ions are chemical entities that carry + or – electrical charges. Hence, positively charged sodium ions are present in very high concentration outside the cell, and low inside. But positively charged potassium ions are the reverse: at high concentration inside the cell and low outside. Because potassium ions are bigger than sodium ions, there is a slight difference in the overall positive charge concentration, i.e., slightly lower on the inside. This results—when the cell is at rest—is an average voltage of about -0.07 volts (70 millivolts) inside versus outside the cell membrane. This can be stable because the soapy (oily) nature of the membrane itself does not allow ions to cross, thus making it a good electrical insulator. However, because the cell membrane is so fluid—always in motion and exchanging various materials between the inside and outside to maintain homeostasis—that there is naturally some leakage that occurs. Therefore there are special "ion pumps" that actively pump sodium ions out and potassium ions in, in order to maintain the electrical and ionic balance.

There are also special pores or channels (also called "transporters") that allow particular ions to diffuse between the inside and outside or vice-versa, to help maintain that 0.07 volt "resting potential." A critical channel that serves this function allows negatively charged chloride ions—which are highly concentrated outside versus inside the cell—to diffuse in, to keep the inside negatively charged relative to the outside. Other channels serve as switches, actually allowing sodium and potassium ions or positively charged calcium ions to diffuse through the membrane, thus throwing the switch on. In fact this is how all nerve impulses—manifesting as sensations and muscular responses—are propagated along the membranes of nerve cells ("neurons"). As for macrophages, the detection of the presence of invading microbes opens calcium channels, and because calcium ions are also positively charged, the membrane voltage goes to zero and the macrophage becomes primed to cause inflammation. But this also happens inappropriately if the resting potential of 70 millivolts deteriorates enough because not enough negatively charged chloride ions diffuse inside to maintain it. In terms of macrophage function, the most important of these regulatory ion channels are what are called "glycine-gated chloride channels," aka "glycine receptors." High concentrations of the free amino acid glycine are required to keep these glycine receptors open, so that the chloride ions can diffuse in at an adequate rate to prevent the macrophage membrane from losing its resting potential inappropriately such as when the cell is actively performing phagocytosis in response to the detection of dead cells or cell debris. The normal extracellular concentration of glycine is established to be about 150-350 micromolar units because that is the average range of concentrations that exist in the general population. The trouble is, even 350 micromolar is not high enough to prevent inflammation when there is only tissue damage but no infection. For that stability, the concentration of free glycine outside the cell (reflected in the blood plasma glycine concentration) needs to be over about 500 micromolar units. The lower it is below that, the more precarious the situation, i.e., the more the macrophages are prone to cause inflammation unnecessarily.

Immunologists will argue that the above description of the inflammation switch is a gross oversimplification. After all, there are many different types of macrophages, specific for different organs in which they are found, and in the specificity of signaling chemicals that activate them. Thus, for example, the white blood cell macrophages called neutrophils are particularly responsive to bacterial infection, whereas the white blood cell macrophages called eosinophils are more reactive to microbial parasites. Beyond activation, there is also a host of different cytokines designed to more efficiently attack different microbes. In the chronic inflammation situation wherein there is no infection present, there are still different kinds of macrophages activated, causing the various different forms of chronic inflammatory disease. Thus, for example, eosinophils are typically over-activated in asthma. Indeed there are many different switches, and the most abundant one is typically the opening of channels in the membrane for the positively charged calcium ions—rather than sodium ions—to come in and activate the many different biochemical pathways needed for an inflammatory response. But remember that it is the inappropriate or excessive activation of macrophages that is the problem in most chronic illnesses—not the normal, appropriate activation in response to infection. What glycine does—via the glycine receptors (glycine-gated chloride channels)—is stabilize the cell membrane against anything that will cause the inappropriate depolarization. Hence we recall the metaphor of the glycine receptor serving not as the universal inflammation switch or trigger, but rather as the universal "trigger lock," preventing any or all of the diverse triggers from being erroneously activated. Importantly, this role provides the basis for the universality of the stabilizing role of glycine vis-à-vis inflammation of any kind.

Chapter Five
Red Flags

In February of 2009, while I was in the midst of both laboratory animal research and self-experimentation on glycine, a group of scientists at the University of Michigan led by Arul Chinnayan (first author: Arun Sreekumar) raised a red flag with the publication of a study on cultured prostate cancer cells in the prominent scientific journal, *Nature*. Their research claimed that the addition of sarcosine (a metabolite of glycine) or glycine to cell cultures of benign human prostate cells made them turn into invasive cancer cells. One of the unfortunate aspects of scientific research publication is the fact that the prominence or high "impact factor" of the particular journal is more determinative of the study's effect on medical research and practice than the actual scientific merit of the published study. As *Nature* is one of the highest impact journals in the life sciences in the world, the Chinnayan group's study had a chilling effect to say the least.

Adding to the negative impact of the study were the difficult rules by which one could attempt to refute the findings by submitting a letter to be published subsequently in the journal. This particular journal required that a letter first be sent privately to the original authors, and that it could only be submitted to the journal for publication following a response or lack thereof to the private letter. I played by all their rules, and got no response from the Chinnaiyan group, and the journal declined to publish my group's detailed letter, which included our own research findings on the biochemical effects of methionine restriction and glycine supplementation. In our letter, we noted that their finding that adding glycine "induced invasion" in benign prostate cells suggested that glycine supplementation might cause or aggravate cancer. We called this suggestion unwarranted, mainly because the so-called "benign" cultured cells they used (RWPE cells) were actually virally transformed cells, i.e., potentially cancerous cells in a non-invasive state. Moreover, the concentrations of glycine to which the cells were exposed were wildly different from actual physiological levels in the body—only about 10% of normal blood levels. But the real proof of the proverbial pudding came three years later, when the Chinnaiyan group (first author: AP Khan) published another study which confirmed the carcinogenic potential of sarcosine in experimental mice, specifically, that the "addition of sarcosine, but not glycine or alanine, induced invasion in RWPE cells..." Curiously they failed to mention—in their 2012 study—that the lack of effect of glycine was ***contrary*** to their earlier result in 2009, when they reported that glycine "induced invasion in these (benign) cells." In fact such a tumor promoting effect of glycine would have been *contrary* to their findings about sarcosine, and the authors should have said so! Hence, their new finding of what might be called an inconvenient truth was studiously avoided in their lengthy 2012 Discussion section. Rather, they just noted that the lack of effect of glycine was "consistent with our (their) *in vitro* data." Hence, the red flag the Chinnayan group had raised in 2009 was most inconspicuously lowered in 2012. I still get occasional calls from customers concerned about taking my glycine supplement, based on the 2009 false alarm.

However, just before the Chinnayan red flag was lowered, yet another, bigger red flag was raised in the most prominent scientific journal *Science*, by a group led by Mohit Jain of Harvard. This time, the (tacit) implication was that glycine supplementation might cause or exacerbate a host of major human cancer types, including breast cancer! The very title of the paper was full of foreboding in terms of glycine supplementation: "Metabolite Profiling Identifies a Key Role for Glycine in Rapid Cancer Cell

Proliferation." Specifically, the Jain group utilized a panel of cultured human cancer cell lines from the collection called the NCI-60; sixty cell lines derived from actual human tumors of nine major cancer cell types including breast, colon, lung, et al. These cultured cell lines had been developed in the 1990s by the US National Cancer Institute, an organ of the US government's Department of Health and Human Services (HHS) and the largest entity funding cancer research in the world. The Jain group analyzed the cultured cells' metabolism by a method known as metabolomics, which enables the measurement of hundreds of metabolic intermediates and products simultaneously. It is one of a family of highly computerized methods including such others as proteomics, genomics and lipomics, that all measure hundreds of substances simultaneously. These are not really scientific disciplines in their own right, but technological tools of science that I would call a mixed blessing. I say this because the proliferation of these tools has resulted in what is known as "data-based science," in contrast to traditional "hypothesis-based science."

In my early days as a scientist—back in the 1970s and 80s—hypothesis-based science was pretty much all science, as generating and testing a hypothesis is a critical step in the scientific process. In data-based science, first one collects all the data possible in a given area of study—because one can—and after computer analysis discerns which differences between groups are statistically significant and which are not—one then thinks up hypotheses to try and explain the findings. Back in the 1970s and 80s, this approach was called a "fishing expedition," and no agency would ever consider funding such an expedition, because the multiplicity of things to be measured would make such a study impossible: One needed to generate a hypothesis first, to delineate a reasonable number of things to measure, in order to test the hypothesis. In fact, the investigations of both the Chinnayan group in 2009 and the Jain group in 2012 were data-based studies utilizing metabolomics.

Hence the very idea that glycine was particularly important in cancerous tumor growth was not arrived at logically, through the study of previous findings and known characteristics of cancer cell growth; rather it just popped out as significant from examining all the possible data. I will have more to say about "-omics"-based research later on in this book, but my summary view is that the "-omics" constitute excellent and unparalleled ways to test hypotheses. But to generate hypotheses; not so much.

Getting back to the work of Jain et al., they not only studied the established cell lines of the NCI-60, but they also studied cells grown from actual human breast cancers, and they were able to correlate enhanced synthesis of glycine in the mitochondria of the tumors (compared to normal cells) with more rapid growth and greater lethality of the tumors. Specifically, they reported "discovering an unexpected increased reliance on glycine metabolism in rapidly proliferating cancer cells" compared to rapidly proliferating normal cells. Hence they suggested that their research might lead to breast cancer therapies that inhibit the production of glycine by cancer cells. So even though they did not address the topic of glycine supplementation *per se*, one can see how the specter of this common nutrient's making cancers grow faster and kill faster cast a pall over the idea of consuming more glycine!

That brings us to the question: What exactly do the Jain group's findings mean? After all, there is no reason to doubt their validity, subject to the obvious limitations related to measuring the activity of cells grown in a test tube, rather than in a living patient's body. The first question I ask is whether the concentrations of glycine that the cells were subjected to are relevant to physiological concentrations. Recall that normal (though not necessarily healthy) levels of glycine in the blood plasma are between about 150 and 350 micromolar units. The Chinnayan group tested their cells at 25-50 micromolar (about 1/10 normal) and the Jain group used 140 micromolar (just below the bottom of the normal range.) In both cases, therefore, the levels of glycine that made the cancer cells grow better were below normal levels. This does not mean their data should be discarded—not at all. Rather it might be useful to test hypotheses concerning glycine and cancer.

Chapter Six
The Evolution of Cancer

It is common knowledge that malignant tumors arise when cells multiply out of control, after mutations (changes in their DNA) accumulate in such a way that a cell is transformed into a parasitic cell that escapes the normal controls of cell division and other regulatory mechanisms. In the process of transformation—which may take years—the mutated cells are subject to a Darwinian natural selection process. During this process, the mutant cells are selected for their ability to multiply rapidly, invade normal tissues and escape the immune system's mechanisms for finding and destroying them. Thus, the "fittest" cells survive to eventually form a "successful" malignant tumor, i.e., a dangerous cancer.

A state of chronic inflammation is now understood to provide an environment that breeds cancer. That is because lots of cellular proliferation is necessary for the healing process; for the building of new tissues to recover from cellular death that follows any sort of injury. But in a state of chronic inflammation, damage keeps being done, so the proliferation of cells for the purpose of tissue repair also becomes chronic. This chronic proliferation multiplies the opportunities for mutations to occur (since cellular proliferation requires the replication of DNA) and for abnormal and potentially cancerous cells to arise.

Earlier, we discussed how a state of chronic inflammation will occur if glycine levels are chronically too low to regulate the macrophages that initiate the inflammatory process. We also saw how a diet rich in muscle meats (and therefore rich in the essential amino acid methionine) and poor in bones and connective tissues (and therefore poor in the amino acid glycine) puts the body into a pro-inflammatory state.

Now the stage is set for abnormal, mutated cells to be generated with increased frequency, and for the mutated cells to be subjected to a natural selection process. Thus, the "fittest" cells would be those that thrive in a high-methionine and low-glycine environment. Both methionine and glycine supply small pieces of the molecules—called purines and pyrimidines—that form DNA, and the formation of lots of DNA is needed for tissue growth and tumor growth. As noted earlier, too much methionine is treated by the body as toxic. In fact, it is known to be toxic in that it inhibits growth. Hence it stands to reason that a more successful tumor cell in a high methionine/low glycine environment would be one that excels at getting rid of excess methionine on the one hand, and at making plenty of its own glycine on the other. Such a cell would thrive in such an environment.

In fact, it has been known for decades that human cancer cells generally are far more dependent on exogenous methionine than are normal cells, the latter having several ways to recycle and salvage methionine molecules. This has even led to the development of cancer treatment strategies that employ a methioninase enzyme that actually eliminates methionine from the bloodstream. Returning to the first study that raised a red flag about glycine—that of the Chinnaiyan group in 2009—we can see that the principal finding of their metabolomic investigation was an accelerated activity of the one pathway that eliminates methionine, an enzyme called GNMT, for glycine-N-methyltransferase. That's why they measured an increase in the production of sarcosine, the derivative of glycine that results from this enzymatic reaction. In the metabolomic study of the Jain group—the second red flag raised about glycine—the main finding was that the most rapidly growing and lethal human cancers had a markedly elevated capacity to make their own glycine. Thus, in actuality, these studies provided evidence in favor of the hypothesis that life-threatening malignancies evolve in precisely the environment of elevated methionine and decreased glycine!

One would think that, since the reciprocal metabolic relationship between glycine and methionine has been documented in the peer-reviewed literature for many years now, scientists in top institutions publishing in top journals would be able to connect the dots. But unfortunately, the hyperspecialization of science is a trend that has only accelerated in recent years. It is as if specialists of different species of trees do not realize they are researching the forest. What can one expect to learn if, when researching a given disease process, one does not consciously relate the findings in one experimental system to other areas of knowledge of human physiology and biochemistry?

Another particularly relevant case in point is a study I mentioned earlier in this book, i.e., the "data based" study by Schmidt et al. from Oxford University, published in 2016. It was part of a very large, ongoing high profile study in Europe (the EPIC study, which stands for The European Prospective Investigation into Cancer and Nutrition) sponsored by the IARC, or International Agency for Research on Cancer of the World Health Organization. EPIC is described by the WHO as "one of the largest cohort studies in the world, with more than half a million

(521 000) participants recruited across ten European countries and followed for almost 15 years." The aim of the Schmidt study was to correlate differences in the amino acid content of the diets of vegans, lactovegetarians, fish-eaters and meat-eaters (omnivores) with differences in the amino acid content of the blood plasma of these four groups of normal men, selected from among UK participants in the EPIC study. Thus, the researchers analyzed the dietary intake of the 392 participants (98 in each dietary group) and recorded their amino acid content, and then measured the free amino acid content of the subjects' blood plasma.

One would think that the researchers—being on the faculty of Oxford, a premier "world class" university, and conducting an arm of such a prestigious multinational study as EPIC—would expect the relationships among the plasma amino acid content and the different dietary amino acid intake to reflect what was known about the interactions of different amino acids in human metabolism. In particular, recall the reciprocal relationship between glycine and methionine noted earlier in this work. Being aware of the fact that excess methionine actually depletes glycine, it should be expected that meat-eaters (omnivores, that is, who normally throw the bones and connective tissues of meat, fish and poultry into the trash instead of the soup), would have lower levels of glycine in their blood plasma than do vegans, even though the high quantities of protein eaten overall by the meat-eaters renders their intake of all the amino acids higher than that of vegans.

But that is not what the Schmidt group hypothesized, for they are "data-based" researchers, with essentially no hypothesis at all. Their aim was therefore simply: "to investigate the differences in plasma concentrations and in intakes of amino acids between male meat-eaters, fish-eaters, vegetarians and vegans in the Oxford arm of the European Prospective Investigation into Cancer and Nutrition". Of course, their analyses of amino acid concentrations were correctly performed, and they found, in their measurements of the blood plasma, "For glycine, vegans had the highest concentration and meat-eaters the lowest."

As part of a proper scientific investigation, the Oxford group thus needed to discuss their findings, including reviewing the relevant literature, to determine the clinical relevance of their statistically significant findings, especially those they found surprising. But in their discussion of results, they invoked the common generalization about "complexity," i.e. "The plasma concentration of amino acids is a result of a complex interplay between dietary intake, (and) tissue breakdown," and simply reported that among the four different dietary intake groups, there were significant differences in the concentrations of six amino acids, and "The largest percentage difference between meat-eaters and vegans in plasma concentration was found for glycine (16%), for which vegans had the highest concentration." All they offered to "in part explain" this was that soybeans—which the vegans ate more of—have lots of glycine, and how higher glycine has been reported to be associated with a lower risk of diabetes. (This was a 2013 finding from another group in Germany, which was also part of the EPIC study team.) But they offered no potential explanation as to why the meat-eaters had the lowest plasma glycine of all, even though they ate the most glycine.

Hence, as disappointing as it was that these presumably erudite authors did not expect to find the big differences between vegans and meat-eaters in terms of methionine and glycine, it was more disappointing to read that they were unable or unwilling to do the basic library research that would have revealed to them the knowledge of the glycine-methionine connection, even after they obtained their surprising results.

Chapter Seven
The State of Life Science Research

These days, one is all too familiar—especially in relation to public health and the COVID pandemic—with the rallying cry "Follow the science!" And within scientific circles, the familiar rallying cry is: "Show me the data!" But as we have just seen in the preceding chapter, the data is not of much use without proper interpretation. Do the data confirm or at least support the underlying hypothesis, or perhaps disprove or mitigate against it? In the absence of a hypothesis, the data can easily be simplistically interpreted to lead one in exactly the wrong direction. Thus, the fact that the most aggressive human cancers make and utilize more glycine has been misinterpreted to mean that glycine supplementation is bad for you, and might cause cancer or make cancer worse. Yet we see that when these findings are interpreted in the broader context of tumor evolution, it makes perfect sense that successful tumors will have to make more glycine than normal cells, if they arise in an environment where glycine is below a healthy level.

If we are therefore to follow the science, it needs to be real science; that is to say whole science: embodying the whole of the scientific method, rather than being just about the data. Hence it is worth considering here just what constitutes the scientific method.

The very word "science" is derived from the Latin verb "scire," meaning "to know." Hence the scientific method is a method for acquiring knowledge about some phenomenon. In particular, science aims to discover the natural law which governs the phenomenon in question, in other words, the "Why?" Thus, the scientific quest always begins with a question. When the mind entertains a question, an answer will usually come to mind, based on previous observations and experiences. That answer would be the hypothesis. But it is, as it were, only a hypothetical answer, because it needs to be tested in the phenomenal (physical) world. Hence the mind is presented with a second question, i.e., how can the hypothesis be tested? In the scientific process, any hypothesis that cannot be tested must be rejected. So the mind seeks a model system—some sort of controlled experiment or set of observations in the phenomenal world; a metaphor for the phenomenon under study—to test the hypothesis. Then comes the actual conducting of the experiment: This is where most of the money is spent, and where the data are generated. If the experiment was designed and conducted properly to answer the original question, the results—the experimental data—should reveal whether the original hypothesis was correct, in whole or in part.

But these days, it's all about the data. And as we have just seen, the same dataset can be interpreted in different ways, and not necessarily correct ones. It has therefore become more and more difficult to sort through the astronomical expansion of life science research data in search of what is correct and important and relevant to answering questions about human health and disease. One scientific principle, the application of which is often forgotten, is typically referred to as "Ockham's Razor." This principle, attributed to the 14th Century theologian William of Ockham, is commonly expressed as a preference for the simplest possible hypothesis that can answer a given question. A familiar historical example would be the Ptolemaic hypothesis that the sun revolves around the earth, as do the planets, in complex gyrations called "epicycles." In stark contrast is the now accepted Kepler model that has the earth and other planets revolving around the sun in simple elliptical orbits.

Yet curiously, findings from disparate areas of study that point to a simple common factor tend to be met with more, rather than less skepticism. The seminal 1999 review by Wheeler et al of the Thurman group in 1999 illustrates this tendency:

"For many, it is difficult to fathom that beneficial effects can be obtained in several disease states with the simplest amino acid, glycine. However, evidence continues to mount in favor of this idea."

No, when disparate findings appear to be connected to a single factor or substance, the problem is actually more easy to fathom, not difficult to fathom! That is the message of Ockham's Razor. One does not reject a hypothesis because it is too simple; rather, because it is too complicated! Yet these days, as I noted earlier, the claim that the problem is complex is often a cover for a lack of knowledge, a hedge to hide behind while maintaining a façade of wisdom and expertise.

But, adding to the tendency to believe that the causes of all manner of dysfunction in complex systems that comprise human physiology must themselves be complex instead of simple, is the tendency to think of anything that normalizes abnormal function to be some kind of drug. Even the recent (2021) review article (from a group headed by Zhending Gan of the South China Agricultural University in Guangzhou) that catalogs the benefits of glycine supplementation in diseases rooted in inflammation, speaks of "the potential application of glycine supplementation as an adjuvant therapy in macrophage-associated diseases." Don't you see? Glycine may be a terrific drug! But just think: these days, no one would ever consider the B vitamin thiamine to be a drug that cures the disease pellagra, nor vitamin C to be a drug that cures the disease scurvy, so why would a single substance that stops inappropriate or excessive inflammation be considered a drug?

An important part of the answer to that question is rooted in the belief that glycine is generally considered "non-essential." So why would anyone place too much credence on the importance of a nutrient we know to be non-essential? In another context, it is interesting to note that, during the recent pandemic of COVID 19, many state governments in the US closed many businesses which were deemed to be "non-essential." Hence many hard-working citizens learned the hard way that their livelihood—certainly essential to their own families' well-being—could be thought of as "non-essential!" So we might say that the word "non-essential" itself is toxic: destroying livelihoods in the workaday world, and throwing scientific research off the trail of finding the root cause of so many illnesses that destroy lives. I have become convinced that these diseases should be thought of as glycine deficiency diseases.

Chapter Eight
Anti-Inflammatory

With the recent realization that inflammation lies at the core of so many modern-day ailments, the term "anti-inflammatory" has become something of a buzzword. People search for foods and supplements and drugs that are "anti-inflammatory" as things that might improve health. That makes perfect sense, and I'm happy to promote the use of supplemental glycine as anti-inflammatory and beneficial, as glycine is the ultimate anti-inflammatory nutrient. But the term "anti-inflammatory" covers quite a bit of territory. There are, for example, supplements marketed to "fight inflammation." This is because these substances actually suppress, in some way, some stage or stages in the multi-stage inflammatory cascade. So for example, the commonly used non-steroidal anti-inflammatory drugs ("NSAIDs, e.g. ibuprofen and acetaminophen) suppress the generation of chemical signals (called prostaglandins) that recruit additional macrophages to the site of inflammation.

More recent types of drugs called "biologics" are actually semi-synthetic antibodies to specific chemical poisons that form part of the inflammatory response. These are used to treat conditions that produce symptoms that are traceable to specific of these inflammatory components. Perhaps the most well known example is adalimumab (the most common brand is Humira®), which has been around for over 20 years. The last three letters "mab" stand for monoclonal antibody. Unlike natural antibodies, "mabs" are products of genetic engineering such that these antibodies do not generate antibodies in the patient that would cause a rejection reaction of some sort. That makes them safe to use on one level, but fundamentally, these "mabs" are dangerous precisely because they are designed to disable a natural function of the immune system.

Adalimumab, for example, specifically inactivates the cytokine called "tumor necrosis factor alpha" (TNFα). Adalimumab is used to treat a variety of chronic inflammatory conditions, including rheumatoid arthritis, ankylosing spondylitis, Crohn's disease, et al. It works because TNFα is near the top of the inflammatory cascade; a product of M1 activated macrophages that can generate fever and induce cell death and all manner of damage typical of inflammation. This makes TNFα an important component of the inflammatory response to infectious disease. That's why, in all the advertising for all the mabs, there is a whole list of warnings about the development of serious, potentially life-threatening infections with use of the mab.

Since they are proteins (All antibodies are large protein molecules), these medications are only available as injectables, because the protein structure would be broken down by the digestive system if taken orally. These medications are also chemically modified to be released into the circulation slowly over time, so dosing needs to be repeated after some weeks or months. Of course, all these drugs do not act to cure any of these inflammatory diseases, precisely because they fight inflammation, rather than naturally regulating it, as adequate glycine does. But remember that glycine—which is not a drug of any kind—only sets the threshold for macrophage activation higher (i.e., to a truly healthy level), so that inflammation can still take place when it is appropriate, i.e., in the presence of an actual microbial infection.

Still, one would think that long term use of an anti-inflammatory medication would lessen the inflammatory damage over time, so that there should be an improvement, provided the absence of natural inflammation did not result in serious infectious disease. That's why Johanna Luitjens and colleagues at the University of California at San Francisco (UCSF) studying the long-term effects of NSAIDS on osteoarthritis in the knee, found their own results to be counterintuitive. They reported their findings at the Annual Meeting of the Radiological Society of North America in 2022. Contrary to their expectations, MRIs of the knees of patients who used these NSAIDS (ibuprofen or acetaminophen) actually evidenced more degenerative changes in their knees than those who did not take NSAIDS. Why? It does seem paradoxical, but in actuality the answer is quite simple. The issue is rooted in imprecise language. Although the NSAIDS act, in part, as anti-inflammatory agents by blunting the recruitment of inflammatory cells (part of a group of drugs known as cyclooxygenase or COX inhibitors), their main immediate pharmacological action is that of an analgesic. Analgesics act by directly reducing pain signals to nerve cells. That's why these drugs are so popular: They reduce pain produced by the movement of a joint which is also suffering inflammation. And although inflammation produces pain, the typical pain of motion in osteoarthritis is not the direct result of inflammation, rather the result of moving a damaged member and directly irritating pain-sensitive nerve endings.

In other words, people take NSAIDS in order to be able to work through the pain by reducing the pain. Therefore, remaining active with joints already damaged by chronic inflammation will cause further damage to the joints, neutralizing or even exceeding the anti-inflammatory benefits of NSAIDS. That is only surprising because we call drugs which are primarily analgesics, anti-inflammatories.

Here, then, is another example of confusion and misdirection resulting from using the wrong word to describe something, rather like calling glycine a "non-essential" nutrient.

Chapter Nine
Back to the Science, a Global Hypothesis Regarding Glycine and Inflammation

The Hypothesis

Whereas most chronic diseases are traceable to some form and degree of chronic inflammation, and other conditions or diseases are traceable to acute inflammation in the absence of any benefit of inflammation, and whereas glycine stabilizes macrophages from inappropriate or excessive priming by permitting adequate influx of chloride ions through glycine-gated chloride channels (aka glycine receptors) on the macrophage cell surface, it is hypothesized that all such disease conditions, whether acute or chronic, result from a deficiency of glycine in the blood plasma and may therefore properly be referred to as glycine deficiency diseases or glycine deficiency conditions.

If this hypothesis is correct, we should expect to make the following observations:

1. The supplementation with glycine of diets of those suffering from any of these conditions will cause a rise in

plasma glycine concentrations and symptomatic improvements in all these conditions.

2. People suffering from any of these diseases or conditions will have plasma glycine concentrations lower than those of unaffected controls.

3. A clear mechanism for the causation of these conditions should be demonstrable, i.e., what caused the stimulus for activation of macrophages that caused the condition, which activation would be putatively prevented by raising plasma glycine concentrations above the threshold needed to prevent such activation. This would include any condition in which the macrophages would be presented with the results of tissue injury, i.e., dead and dying cells and/or cellular debris. An obvious example would be blunt injury, where tissue damage clearly occurs. Less obvious are recurring physiological situations in which cell death normally occurs, such as ovulation in the ovaries, or premenstrual regression of breast tissue during each non-conceptive monthly cycle and during weaning. The same is true for some normal developmental programs, such as in the brain, where embryonic or early childhood brain tissue is resorbed as the adult brain is being formed. These conditions will be examined in some detail later on in this book.

4. Glycine deficiency diseases or conditions should be more common among people who consume less glycine and/or more foods which deplete plasma glycine.

5. Since all these chronic putative glycine deficiency conditions have become more common in recent decades, dietary habits in the population at large should be known to have shifted, in recent decades, toward less glycine intake and/or greater intake of methionine-rich foods, such that plasma glycine levels in the population at large have decreased over the last few decades.

The Proof

These days, proof from an actual clinical trial must be produced in order to establish the role of any substance in normal or abnormal physiology—in other words, almost that of a controlled experiment is required. The best type of clinical trial—one that approaches the design of a real controlled experiment—is a placebo-controlled, randomized clinical trial, and the bigger the better, as the results become more statistically reliable as the numbers increase. Yet, while it is generally true that a placebo-controlled randomized clinical produces more reliable results than studies in which many factors cannot be controlled, convincing proof of anything can be provided by other means.

For example, the weakest sort of study is a correlational study. In a correlational study, a population of observed subjects with and without a given disease is compared to, say, the overall exposure to the putative causative agent in the general population. In such a study, one is not even comparing, on an individual basis, those with versus those without the exposure in question that developed or did not develop the disease in question.

Nevertheless, in the early 2000s, a study of the correlation between the sudden drop in breast cancer incidence in the general population of US women over 50 years of age, and the sudden drop in combination hormone replacement therapy (HRT) use in postmenopausal US women, was taken to be decisive. That is, it was taken as legitimate proof that combination HRT is a causative agent in breast cancer. Importantly, the known cellular proliferation-inducing effect of these drugs had also been established, so that a likely mechanism for the cancer-causing effect was present. It was in fact a placebo-controlled, randomized clinical trial of combination HRT as a putative preventative for heart disease—called the Women's Health Initiative or WHI trial—that led to the sudden drop in combination HRT use. Specifically, interim results after 2.5 years of the five-year study showed an increase—

rather than the expected decrease—in heart attacks, forcing the early termination of the WHI study. The attendant publicity included the fact that breast cancer incidence was also increased among WHI study participants, which resulted in a 70% decrease in combination HRT prescriptions the following year. The findings of the WHI study have been recently reviewed by Petra Stute of the University of Bern, Switzerland.

In the case of the present hypothesis concerning glycine, there is a wealth of study data from a multiplicity of types of studies, all of which point in the direction of confirmation of the hypothesis. Clinical trial data, however, are rare. In the remainder of this chapter, we'll examine the evidence that fulfills the first two elements of proof listed above.

1. Glycine supplementation showing improvement in an inflammation-based condition:

One clinical trial by Cruz et al.—apparently a one-off—was published in 2008. It was in fact a randomized, placebo controlled trial of oral glycine, 15 grams per day, in type 2 diabetic patients in Mexico City. Although the study was small by clinical trial standards—74 patients—there was a significant reduction in the premier blood test marker for type two diabetes: Hemoglobin A1c, which was reduced, on average in the glycine-treated group, from 8.3 to 6.9. This is quite a dramatic result, all the more remarkable because the patients were consuming their normal anti-diabetic medications all along. (It would not be an ethical study otherwise.)

In their discussion of results, Cruz et al. focused on the results that showed a decrease in inflammatory markers in the blood with glycine treatment, and cited research from the early 2000s which linked type 2 diabetes to inflammation. Unfortunately, the authors did not highlight the dramatic reduction in Hemoglobin A1c they also observed. Were it up to me, I would have entitled the paper: "Glycine supplementation reverses type 2 diabetes." Moreover, the study was not published in one of the top "high impact" endocrinology journals, and it remains in relative obscurity to this day.

The Cruz study is one of the very few clinical trials to date to focus on any anti-inflammatory effects of glycine supplementation. (There have been other trials of glycine supplementation in psychological disorders, based on glycine's action as an inhibitory neurotransmitter in the brain, and without striking results.)

2. Comparison of blood glycine levels in subjects with versus without an inflammation-based condition:

This is an area in which metabolomics—data-based science—is most useful. In terms of clinical studies, it is the easiest kind to perform. Often, there are blood samples that have already been collected from many—sometimes tens of thousands—of patients, and literally hundreds of metabolites may be measured in a single pass from a very small amount of stored plasma or serum.

With respect to type 2 diabetes, if it is indeed a disease of glycine deficiency, differences in glycine concentration should be demonstrable in matched groups of subjects with versus without type 2 diabetes in a metabolomic analysis. Indeed, the high and increasing prevalence of this condition has focused a great deal of research attention on it. Exemplifying this trend, Harvard nutritional scientist Marta Guasch-Ferré and coworkers published a systematic review and meta-analysis (SRMA) in 2016, which pooled results from 46 separate studies where metabolomics had been used to measure a wide spectrum of metabolites, including all the amino acids. Only two of the amino acids—

glycine and glutamine—were found to be at significantly lower concentrations in the blood of pre-diabetics (a condition typically preceding onset of type 2 diabetes and characterized by insulin resistance) and type 2 diabetics. This study of studies—comprising thousands of patients—does not prove the glycine hypothesis in terms of causation of diabetes, but failure to find this result—an inverse association between glycine and diabetes or prediabetes—would essentially disprove the glycine hypothesis. After all, how could you view diabetes as a glycine deficiency disease if there is no difference between blood glycine concentrations among diabetics versus nondiabetics? Finding that glycine levels are lower in diabetics and pre diabetics than in a healthy, matched population therefore provides essential—though not sufficient—support for the glycine hypothesis.

The Guasch-Ferré study is important, although it suffers from its being a metabolomics study. That is, as a data-based study, it does not begin with a hypothesis concerning why one might expect particular metabolites to be increased or decreased in diabetics. Rather it is indeed a fishing expedition. The findings of significantly decreased plasma glycine and glutamine were significant "fish" that were caught. But what did it mean? This was explored in a very cursory fashion in the Discussion section of the paper. No specific hypothesis was offered; rather just the citation of a 2002 study showing generic effects of amino acids on glucose metabolism. To the data-based scientist, hypothesis comes as an afterthought—with most of the paper devoted to statistical analysis of data.

Remember that the Guasch-Ferré study was published four years after the publication of the Cruz clinical trial of glycine for type 2 diabetics, and both the significant reduction of Hemoglobin A1C and of markers of inflammation were among the positive effects of supplemental dietary glycine that the Cruz study documented, and the anti-inflammatory effects of glycine were demonstrated almost 20 years after the pioneering work of the Thurman group. One might expect, therefore, that Guasch-Ferré et al. might have connected the proverbial dots: diabetes, inflammation and glycine. Then again, a look at the stated objective of the Guasch-Ferré paper reveals the self-imposed limitations on the group's efforts:

OBJECTIVE

To conduct a systematic review of cross-sectional and prospective human studies evaluating metabolite markers identified using high-throughput metabolomics techniques on prediabetes and type 2 diabetes."

Where is the objective to further understand diabetes, what causes it, how to predict it and cure it? These objectives may seem implicit merely by virtue of the fact that the study was funded by the NIH and the American Diabetes Association. Instead, the fixation on the data obscured the importance of the group's own findings!

I should also mention the connection that the Guasch-Ferré group found with the amino acid glutamine being reduced among diabetics and pre-diabetics. The likely reason is straightforward. Unlike most amino acids, which are made by the liver, the amino acid glutamine is mainly made by muscle. Insulin resistance manifests mainly as a decreased ability of muscle to take up the fuel glucose from the circulation, thus muscle function is inhibited, and a decrease in glutamine production is to be expected. Therefore, we may infer from the Guasch-Ferré group's findings that the decrease in glycine reflects the cause of insulin resistance, and the decrease in glutamine reflects the effect of insulin resistance. (As we shall see later on, the synthesis of glycine is also affected by insulin resistance and diabetes. Hence, low glycine may indicate both cause and effect.) Speaking of other amino acids, another study by a group at Baylor College in Houston, Texas, published in 2011—a clinical trial, actually—showed similar benefits in type 2 diabetes after supplementation with glycine in combination with another amino acid: cysteine. More on that study in a later chapter.

Data-based metabolomic studies have also linked blood glycine levels to other serious chronic diseases linked to inflammation, such as cardiovascular disease. As the glycine hypothesis would predict, such studies also show decreased blood levels of glycine linked to adverse cardiovascular events, mainly heart attacks and strokes.

In 2015 Hartiala and colleagues at the Cleveland clinic used metabolomics to identify genetic variations for genes that code for enzymes involved in the synthesis and metabolism of glycine, as some of these had been identified with reductions in heart disease risk. They discovered a highly significant connection between defective glycine clearance and a 12% lower risk of coronary artery disease (CAD) in women. While this did not prove that the elevated blood glycine levels were responsible for the lower risk of CAD, the authors cited earlier (2012) research that had shown the inflammation reducing and "cardioprotective" effects of glycine. Importantly, they also showed that the significant risk reduction in the genetic variant population disappeared when they statistically adjusted for the increased glycine. This is good statistical evidence that the increased glycine levels in the variant population were responsible for the decrease in CAD risk.

The following year (2016), Yupeng Ding and colleagues at the University of Bergen, Norway, and the University of Florida, looked specifically for a protective effect of glycine for CAD (real hypothesis-based science!). They included over 4,000 participants in their study of male and female patients with suspected stable angina, which is essentially, quiescent CAD. The researchers followed up the participants for an average of over seven years, and compared blood glycine levels of those who had an acute myocardial infarction (AMI, i.e., a heart attack) with those who did not. What they found was about the same risk reduction for AMI (11%, and statistically significant) as the Cleveland clinic had found the prior year for CAD.

Another, more recent observational study on obese teenagers adds support to the importance of glycine deficiency as a cause of cardiovascular disease (CVD). Wagner

Luiz do Prado of the California State University and colleagues at children's health centers in Jacksonville, Florida, Wilmington, Delaware and the Mayo Clinic in Rochester, Minnesota, studied patients aged 14-18 years, to follow up on the growing body of research linking CVD to low levels of glycine. Specifically, since CVD occurs almost exclusively in older adults, they hypothesized that the inverse correlations between glycine and obesity, prediabetes and markers of inflammation (C-reactive protein or CRP, and Interleukin 6 or IL-6) seen in adults, would also show up in children. They also noted that recent studies had shown a distinct decrease in blood glycine levels among obese children and adolescents. Confirming their hypothesis about parallels between CVD biomarkers in children and adults, the authors concluded: "Given that CVD progression is a continuum and the disease itself is not present in children and biomarkers are typically used to monitor CVD in children, the links between GLY and biomarkers of CVD provide evidence for the first time of a potential role for GLY in CVD in children with obesity."

Finally, regarding the work of do Prado and colleagues, it is noteworthy that they are among only a handful of researchers who consider glycine not "non-essential," but rather as a "conditionally essential" amino acid. In my view, that in itself represents major progress!

We will take a closer look at obesity and its consequences vis-à-vis glycine in the next chapter.

Chapter Ten

How Does Glycine Deficiency Result in Specific Disease States?

As we have seen, glycine deficiency results in excess or inappropriate inflammation when there is tissue damage (i.e., cell death). Harking back to Chapter Nine and the glycine deficiency hypothesis I have advanced, any disease state that is rooted in chronic inflammation should be shown to have, as a chronic condition that leads to disease, a chronic situation of cell death in a particular tissue or organ. Here we will discuss in some detail the connection between chronic cell death and the ultimate development of a serious disease state, in most of the diseases that make people sick and die these days.

The types of situations involving cell death can be divided into 4 types:

1. Normal development of certain organs and tissues
2. Certain physiologically cyclical processes.
3. Trauma from any sort of injury, or 'micro injuries' due to normal internal or external physical activity or reinjury from disturbance of scar tissue from

previously healed traumatic injury.

4. Direct harm to normal cells by infectious microbes (to which some level of inflammation is an appropriate response, but overreaction to which produces more serious or chronic disease).
5. In addition to all these conditions producing some level of excess inflammation when glycine is deficient, inflammation may be provoked for unknown reasons, such as in migraine headaches.

1. Developmental Processes: Autism (most commonly now referred to as Autism Spectrum Disorder or ASD)—a serious condition which starts before birth. The development of the human brain—both before and after birth, until about age 25—occurs via the multiplication of precursor cells into solid masses of cells, followed by the elimination of cells that are not needed. These unnecessary cells are eliminated by cells called microglia—the macrophages of the brain—doing their normal phagocytosis of cells which die. But if glycine is deficient, the death of the unnecessary neuron precursor cells primes the microglia to cause inflammation. In 2011, PH Patterson observed "an ongoing, hyperresponsive inflammatory-like state in many young as well as adult autism subjects." Using postmortem brain specimens from ASD v normal subjects, Lee et al observed "a shift in microglial phenotype that may indicate impaired synaptic plasticity and a chronic vulnerability to exaggerated immune responses." More specifically, they described "a significant increase in primed microglia (i.e., M1 v M2 phenotype) in gray matter of ASD compared to typically developing individuals." Although Patterson did not measure nor implicate glycine deficiency, Wheeler, et al. of the Thurman group at UNC had elucidated the role of glycine in macrophage activation back in the 1990s. Moreover, in 2017, Mori et al demonstrated experimentally the ability of glycine to ameliorate brain damage in rats. My own summary analysis of autism as a glycine deficiency condition was published as an e-letter on the British Medical Journal's website in 2018.

2. Physiological cyclical processes: Menstrual cycle

a) Ovulation and ovarian cancer: While ovulation is the normal process of the ejection of a mature egg (secondary oocyte) from alternate ovaries during a woman's fertile years, it is a rather violent process, causing injury to the surface of the ovary. Over a half century ago, in 1971, MF Fathalla at Assiut University in Egypt observed the epidemiological (statistical) association between number of lifetime ovulations and the frequency of ovarian epithelial cancer (the most common type of ovarian cancer in women). He noted that most ovulations in modern times served no purpose, and that the repeated trauma to the ovarian surface somehow promoted the development of cancer. Ovulation, after all, is naturally stopped when a woman is pregnant or breastfeeding, and these situations had become rare by the late 20th century. Artificially, birth control pills (which are generally composed of synthetic estrogen and progestin drugs) generally act via suppression of ovulation, and their use is also known to be associated with a decreased risk of ovarian cancer. However, for almost 30 years the dominant theories as to the cause of ovarian cancer centered on hormones, especially estrogen, but no conclusive connections were found. Finally, in 1999, RB Ness and C Cottreau at the University of Pittsburgh, suggested that "inflammation is a pathophysiologic contributor to the development of ovarian cancer," due to the effects of inflammation causing "cell damage, oxidative stress, and elevations of cytokines and prostaglandins, all of which may be mutagenic," i.e., can cause mutations in the cellular DNA. Meanwhile, any suggestion as to a connection between ovarian cancer and glycine status has yet to appear in the medical literature, to my knowledge.

b) Breast Cancer. Most people think of a woman's menstrual cycle as a process that happens in the ovaries and the uterus. However, the cyclical hormonal changes—which occur in the ovaries—have effects body-wide, including the breasts. Many women experience breast tenderness in the third week of the menstrual cycle, because the combination of estrogen and progesterone that peak during that time cause a growth spurt in the breasts. That is, they cause a burst in cellular reproduction to begin the massive breast growth that will be needed if conception occurs and a baby develops who will need to be fed after birth by the breasts. However, if conception does not occur, all those extra cells that were created need to die off and their substance resorbed, a perfectly normal regression process that happens before menstruation. But if glycine is deficient, this regression process is characterized by a low-grade inflammation. In terms of epidemiology, the science that statistically relates specific exposures to specific outcomes, breast cancer risk is proportional to the lifetime number of menstrual cycles. However, the use of birth control pills is known to increase the risk of breast cancer, in contrast to decreasing the risk of ovarian cancer. That is attributable to the fact that the combination of estrogen and progestin drugs in birth control pills, stimulate cellular reproduction in the breasts for most of the monthly cycle, rather than the normal estrogen-progesterone-mediated growth stimulus that happens only for several days (during what is called the luteal phase) of a normal menstrual cycle. Hence, with more growth in the breasts each month, there is more cell death (tissue resorption), and, if glycine is deficient, more inflammation happening each month.

In addition to the monthly cycle of growth and regression that occurs during non-conceptive menstrual cycles, The growth of breast tissue is much more massive if conception occurs. Then, the hormonally driven growth spurt that happens during the luteal phase of the cycle accelerates with exponential increases in the same hormones (estradiol, the main form of estrogen, and progesterone) secreted by the ovaries and later, by the placenta. That is why the breasts typically double in size during a normal pregnancy. If the hormonal support for this massive growth is suddenly stopped by an abortion of the pregnancy, the resulting inflammation (if glycine is deficient) shows up epidemiologically as an increased risk for breast cancer among women who have had any induced abortions. (I have published, along with several colleagues, systematic reviews and meta-analyses of the research on this subject in 1996 and 2018.) If, however, the pregnancy continues to term, breast cancer risk is known to decrease, compared to no pregnancy at all. This protective effect of full-term pregnancy was first established in 1970 by a landmark World Health Organization (WHO) study led by Brian MacMahon at Harvard. The protective effect is attributable to the differentiation or maturation of the excess breast tissue generated by the pregnancy, into mature tissue capable of producing milk. The terminally differentiated cells of these breasts, being no longer capable of cellular division, thus produce no mutations in their DNA from cellular division, and therefore do not give rise to malignancies. However, there is a measurable increase in breast cancer risk after birth, if the infant is not breast fed, or after weaning if the infant is breastfed. That's because the massive regression of the excess breast tissue produces substantial inflammation, if glycine is deficient. Finally, it should be noted that miscarriage (aka spontaneous abortion) does not produce an increase in breast cancer risk. That is because most miscarriages occur when the embryo or fetus dies too early during the pregnancy for the pregnancy to have generated a massive growth spurt.

3. Microinjuries from normal physical or physiological activity

a) Blood circulation and cardiovascular disease.

The arterial part of the circulatory system constitutes a high-pressure fluid distribution system. As such, there is naturally some level of turbulence, which is maximal around high-pressure branch points. The most turbulence is therefore to be found where the pressure is greatest, where the coronary arteries (which serve the heart muscle itself) branch off the ascending aorta, the carotid arteries (which serve the brain) branch off the brachiocephalic artery, and the renal arteries (which serve the kidneys) branch off the descending abdominal aorta. Not surprisingly then, the major—often life-threatening—adverse events in cardiovascular disease involve blockages that occur at these branch points, i.e., heart attacks, strokes and kidney failure, respectively. It has long been recognized that these events follow years of buildup of atherosclerotic plaques from a chronic inflammatory process, which progressively occlude the arterial opening. Finally, the adverse event often occurs when a blood clot inappropriately forms at this narrowed point, thus abruptly stopping the blood flow at that point altogether. More recent research has begun to tie adverse cardiovascular outcomes with lower levels of blood glycine (as we saw in the previous chapter) with the work of Hartiala and Ding in 2015 and 2016, respectively. Hopefully, researchers will increase their focus on glycine. It should be noted that even the observation of an inverse correlation between adverse cardiovascular events and blood glycine is still restricted to the "normal" range of glycine concentrations, which is really quite sub-optimal. Hence, observed differences in outcomes and glycine levels, while statistically significant, will hardly be as dramatic as one would expect between those with healthy high glycine levels and the general public. This sort of conclusive finding would be expected only in a long-term clinical trial of those with v without glycine supplementation.

To this point I would also hypothesize that glycine supplementation might also serve to dramatically reduce the risk of adverse cardiovascular events due to the inappropriate formation of blood clots (i.e., heart attacks and strokes) among those with existing extensive cardiovascular disease, i.e, substantial occlusion of critical arteries. That is because glycine acts in three important ways to reduce the risk of inappropriate clot formation, as there are essentially two interdependent systems to form clots in the bloodstream: 1)The formation of clumps of platelets (aka thrombocytes, which literally means "clot cells"), and 2) the chemical formation of glycoprotein nanofibers (called fibrin), which turn the liquid blood plasma into a gel.

Normalization of platelet aggregation: In 2013 Peter Schemmer and coworkers of the Thurman group at UNC established the presence of glycine receptors in rat and human platelets, and showed that glycine reduced platelet aggregation (clot formation) in a dose-dependent manner. That means, the more glycine, the less likely were the platelets to stick together. Importantly, the platelets could still aggregate normally—as in response to an actual cut in the blood vessel wall—with concentrations of glycine 2 or 3 times higher than the "normal" (but actually suboptimal) levels.

Normalization of chemical clot formation: When there is damage to the blood vessel wall, the cells of the wall's lining (called the endothelium) release a protein called "tissue factor" which begins a chemical cascade (the coagulation cascade) that amplifies the reaction and which culminates in the formation of a fibrin clot. That is, the last step in the cascade is the conversion (via the activated enzyme called thrombin) of the soluble glycoprotein called fibrinogen into an insoluble matrix; a nano-meshwork, as it were, that turns the liquid blood plasma into a gel. Of course, such a biochemical system that turns a very small injury into a sizable clot capable of plugging the hole in the blood vessel, must be very tightly controlled, lest the whole circulatory system form one big clot, which would be rapidly fatal! Therefore, the clotting system is designed such that, when the coagulation system is activated, another cascade (the fibrinolytic cascade) is simultaneously activated. The fibrinolytic cascade culminates in the enzymatic digestion of the fibrin (via the activated enzyme called plasmin), thus dissolving the clot. However, in the location of the actual injury site through which blood is flowing out of the vessel, another enzyme called factor XIII or fibrin stabilizing factor, chemically cross-links the nanofibers of fibrin, so that the plasmin cannot dissolve the plug that fills the hole.

The two ways in which glycine stabilizes the chemical coagulation system are through normalizing the changes in blood chemistry that result from inflammation, namely:

1) Increased production and secretion of Factor VII by the liver. Factor VII is a circulating protein that combines with the tissue factor (the top of the coagulation cascade) to initiate the coagulation cascade. When there is more factor VII around, the blood clots more readily.

2) Increased production, by the platelets, of another soluble protein called Plasminogen Activator Inhibitor-1 or PAI-1. PAI-1 has been known since 1986 to be part of the inflammatory cascade as well as the fibrinolytic cascade. That is, the penultimate step in the fibrinolytic (clot dissolving) cascade, i.e., the activation, by Tissue Plasminogen Activator (TPA) of the pro-enzyme plasminogen to form active plasmin, is normally inhibited by PAI-1. When there is inflammation present, the platelets secrete more PAI-1, thus rendering clots more stable, and thus more liable to create a blockage (thrombosis). In 2005, DE Vaughan at Vanderbilt University in Tennessee, suggested that PAI-1 was a key player in the formation of blood clots in heart attack and stroke. Hence, a state of chronic inflammation can increase the likelihood of inappropriate clot formation. That is why patients suffering from cardiovascular disease typically have blood that is hypercoagulable, i.e., more likely to form inappropriate clots, and why they are generally prescribed some sort of "blood

thinners," drugs that inhibit coagulation. Aspirin, for example, which inhibits platelet aggregation, is typically recommended for older individuals with any family or personal history of stroke or heart attack or other condition involving inappropriate blood clotting. Prescription drugs, such as warfarin (which inhibits the liver's production of coagulation factors including Factor VII and prothrombin, and Clopidogrel (Plavix®) which inhibits platelet aggregation are commonly prescribed as "blood thinners." Theoretically, elevation of glycine to a healthy level, by glycine supplementation, should dramatically decrease the risk of stroke and heart attack, and/or recurrence of stroke and heart attack, without the use of drugs. While this is hypothetical at this point for patients who need to have their blood's hypercoagulability reduced, it is certainly harmless for anyone to correct a simple nutritional deficiency, regardless of what drugs they might need. Certainly the observational studies of Hartiala and coworkers at the Cleveland Clinic and Ding and coworkers in Norway point in this direction.

b) Pre-diabetes and Diabetes: Diabetes (specifically, Type 2 diabetes mellitus) is not usually ascribed to any sort of injury, but there is reason to hypothesize that this is the case at the cellular level. Metabolic syndrome (aka insulin resistance or prediabetes) and diabetes have historically been ascribed to chronic excess carbohydrates in the diet, eventually exhausting the beta cells of the pancreas from the strain of producing insulin to remove the excess glucose from the blood circulation. However, since at least the 1970s, Prediabetes and diabetes were found to be associated with obesity, specifically, abdominal obesity. Moreover, the causal link seems very clear, since it is now known that losing the extra 20 or 30 pounds or more by diabetic patients actually can reverse the condition. More recently, researchers have observed an association between diabetes and markers of inflammation.

In fact, the secretion of inflammatory mediators—cytokines known to mediate inflammation—by abdominal adipose tissue has been known for over 30 years. In 1993, GS Hotamisligil and coworkers at the Dana-Farber Cancer Institute in Boston observed the secretion of tumor necrosis factor alpha (TNF-alpha, an important mediator of inflammation secreted by macrophages) by adipose tissue in laboratory rodents. They also observed that neutralization of the TNF-alpha enhanced the sensitivity of muscle to insulin, implicating inflammation as a causative factor in the development of insulin resistance and diabetes. But it was not until a decade later that Stuart Weisberg and colleagues at Columbia University in New York City established that the secretion of TNF alpha and other inflammatory mediators including interleukin-6 (IL-6) was not from adipocytes (fat cells) in the adipose tissue of mice and people, but from macrophages embedded in the fat tissue. This was possible because of technological advancements in the detection of cell surface markers that can differentiate between different types of cells and their origins. Like all macrophages, the adipose tissue macrophages are immune cells which arise in the bone marrow and carry the cell surface marker called CD68 (F4/80 in mice).

On the cellular level, Weisberg and colleagues described what has come to be known as a "crown-like" structure. This is the appearance of a large, round, adipose (fat) cell that appears to be surrounded by a crown-shaped form. This "crown" is actually a cluster of macrophages involved in the consumption and reabsorption of the adipose cell which has been damaged in some way. These "crown-like structures" are now known to be a hallmark of local inflammation in adipose tissue.

In 2010, Naim Alkhouri and colleagues at the Cleveland Clinic suggested that apoptosis (cell death; although, as I have described earlier, the adipocyte cell death going on here is more like necrosis or pyroptosis; i.e., inflammatory-type cell death) of adipocytes is the first step in the sequence of events leading to inflammation in adipose tissue, insulin resistance (prediabetes) and type 2 diabetes. The Weisberg group and others had previously demonstrated a direct correlation between adiposity (percentage of body mass composed of adipose tissue) and the average size of adipose cells. Weisberg et al. had also shown a direct correlation between adiposity and the percentage of adipose tissue composed of macrophages, ranging from about 10% in lean individuals to about 40% in massively obese individuals. Hence adiposity, adipocyte cell volume, percentage of macrophages in adipose tissue and measures of inflammation all correlate together. Larger adipose cells (i.e., adipocytes) are therefore more vulnerable to stresses which can induce cell death, and so there is more of it. This attracts macrophages to the tissue to devour the dead and dying adipocytes. If glycine is deficient, the macrophages initiate the inflammatory response which creates more cell death, and the vicious cycle escalates.

While those researching the connection between diabetes and inflammation noted above seem to have been unaware of any role of glycine, the relevance of glycine was not lost on Cruz and coworkers in Mexico City, who conducted a clinical trial of glycine supplementation in type 2 diabetes in 2008. Specifically, citing earlier work by Gonzalez-Ortiz et al. that showed some benefit of glycine in glucose control, and the work of the Thurman group at UNC which revealed the glycine-mediated reduction of the secretion of inflammatory mediators (TNFalpha in particular) by monocytes (white blood cells which are precursors to macrophages), Cruz et al focused on markers of inflammation in their clinical trial, which I summarized in the previous chapter. Among the dramatic findings in their study was a result directly related to the question of adiposity. Specifically, the reduction in hemoglobin A1C in the glycine treatment group—i.e., the reversal of type 2 diabetes in a patient group that was moderately overweight (BMI between 28 and 29)—was accomplished with no change in adiposity (represented by the BMI). Hence, the connection between adiposity, cell death, inflammation, type 2 diabetes and glycine has been clear since at least 2010. Yet it remains largely unknown!

Yet there is something incomplete in the story about how abdominal adiposity causes type 2 diabetes. That is, knowing that the inflammatory cytokine TNFα causes insulin resistance in muscle and fat, and thus makes the beta islet cells—the cells in the islets of Langerhans in the pancreas that actually make insulin—work harder, does not really explain exactly why the beta cells should ultimately fail to make insulin at all and eventually die. After all, there are many bodily cells and organs that suffer from chronic inflammation, but they don't all fail in relatively short order like the beta cells do in type 2 diabetes (or even sooner in Type 1, which is juvenile onset diabetes).

With so much research attention focused on type 2 diabetes, many details of this process have been elucidated over the last two decades. This research was recently reviewed by Wei Ying and coworkers at the University of California, San Diego. It is now known that the islets of Langerhans in the pancreas have lots of resident macrophages, and that the numbers of these macrophages increase with adiposity (obesity). As adiposity increases, not only the number, but also the activation from the non-inflammatory "mend" state (M2), to an active inflammatory "rend" state (M1) increases, with the increased secretion of cytokines which are pro-inflammatory mediators, especially TNFalpha and Interleukin-1beta (IL-1β). Moreover, these cytokines, depending upon conditions which are not so well known, sometimes increase the number and function of beta cells (remember, these are the cells that actually make the insulin and which malfunction in diabetes) and sometimes decrease their function and promote their apoptosis (cellular demise). Whatever the details, the cellular and molecular events that happen with increasing adiposity almost invariably lead to beta cell dysfunction and death and therefore, diabetes. So with all that research, we are left with more questions than answers, and a situation that seems—you guessed it—increasingly complex! Ying et al throw up their hands with such statements as:

"Macrophages can adapt to local environmental cues, and it is possible that intra-islet resident macrophage proliferation is an adaptive response to pathophysiological stimuli. However, the factors that mediate this adaptation remain to be defined."

As the authors say at the end of this review exercise:

"While much has been learned about macrophage-beta cell interactions, many important questions remain unresolved. What are the triggering events with respect to islet macrophage proliferation and expansion? Are initiating signals derived from the beta-cell or are they extrinsic to the islet?" These are certainly good questions. And you might think that, since this is all about a common and serious metabolic dysfunction centered on how the body handles certain nutrients—especially glucose (sugar) and amino acids—that researchers would be looking to dietary components; to the role of specific nutrients, for answers. While there have been studies on fats, studies which implicate saturated fats as contributing to islet inflammation in diabetes, there has been relatively little effort made in studying the role of amino acids in this process, even though amino acids figure strongly in overall islet function in terms of hormone secretion (especially insulin and glucagon).

Thankfully, there has been some excellent research focused on amino acids and their role in obesity and insulin resistance (pre-diabetes) done by Hong Chang Tan at the Singapore General Hospital and colleagues at Baylor College of Medicine in Houston, Texas. Studying a group of patients who underwent bariatric surgery for their obesity, they have been following up on observations of differences in amino acid profiles that were enabled by the ability to measure most (16) of them (as well as over 100 intermediary metabolites and hormones) in a single test, using the relatively new tool of metabolomics we discussed here earlier. Thus, back in 2009, Christopher Newgard and colleagues at Duke University Medical Center in North Carolina showed that eight of the amino acids (including the branched chain amino acids leucine, isoleucine and valine) were dramatically elevated in the blood serum of obese subjects, compared to seven amino acids that were not different from non-obese control subjects. In stark contrast, only the concentration of glycine was dramatically reduced in the obese subjects. These findings raised the question of causal sequence, that is, were the differences in the amino acid concentrations cause or result of the insulin resistance?

In 2022, when the Tan group in Singapore measured the glycine levels of patients before, and six months after bariatric surgery, they found that, in addition to the predicted decrease in obesity and insulin resistance, the glycine levels went up to normal (i.e., the same level as non-obese control subjects) from being 19% below normal. Importantly, Tan and colleagues also did specific biochemical testing on their subjects, and they found that the rate at which the bodies of the obese patients synthesized glycine was 30% below normal. Six months after the surgery, not only had their glycine levels come up to normal, but so did the glycine synthesis rate. In other words, the Tan group was able to conclude that the low levels of glycine found in obese subjects was because of decreased synthesis, and not from an increased rate of glycine metabolism or decrease in dietary glycine consumption.

Based on these clear findings, Tan and colleagues argue that the causal chain suggested by their results is: obesity → insulin resistance → decreased glycine synthesis → hypoglycinemia (decreased blood glycine levels). But this assumes that the glycine levels (and glycine synthesis rate) are truly healthy in the non-obese subjects. If we assume, on the other hand, that "normal" glycine levels are low to begin with (not due to inadequate glycine synthesis, but inadequate dietary intake), then we would have obesity in the presence of hypoglycinemia → insulin resistance and decreased glycine synthesis. This is what I conclude based on the clinical trial of the Cruz group in Mexico city in 2008, which we discussed here earlier. They tested moderately overweight subjects with type 2 diabetes (which follows on insulin resistance and characterized by extreme insulin resistance) not with bariatric surgery, but with 15 g per day of supplemental dietary glycine. Their inflammatory and diabetic markers declined toward normal after three months of treatment, but with no loss of weight. Although their HOMA-IR, a standard measurement of insulin resistance, did not come down to normal it was reduced toward normal by almost ten percent.

I would therefore argue that the initial, diet-associated glycine deficiency was the underlying cause of the metabolic disorder (insulin resistance), with obesity being a secondary cause by generating systemic inflammation in a hypoglycinemic (low glycine) environment. What Tan and colleagues showed, then, was that obesity makes hypoglycinemia worse by decreasing glycine synthesis.

Glutamine is perhaps foremost among amino acids that have been studied in relation to diabetes and specifically to the function of macrophages and inflammation in the development of diabetes. In the Guasch-Ferré meta-analysis discussed in Chapter Nine, you will recall that both blood glycine and glutamine were found to be low in type 2 diabetics. More recent research on glutamine and macrophage activation in regards to obesity and diabetes has been extensively reviewed by Wenkai Ren and colleagues of the South China Agricultural University in Guangzhou. They noted that glutamine is critical to the activation of macrophages to the M2, or cleanup function, as opposed to the M1 or inflammatory function, which is ascendant in obesity and diabetes. Moreover, they traced the biochemical pathways implicated in this role, and noted glutamine supplementation might be a useful strategy in the treatment of diabetes, as it has been found to be beneficial in laboratory experiments.

However, they also noted that (contrary to most of the amino acids, including glycine) glutamine is mainly synthesized in muscles, whose function is compromised in obesity and diabetes. Therefore, they admit that glutamine might well be low as a result of the diabetes, rather than a cause. This is reminiscent of the recent work of the Tan group, which also showed a decrease in glycine synthesis as a result of the metabolic disorder (insulin resistance). As for glutamine, in the typical, muscle meat-rich diet consumed these days in the industrialized world, glutamine consumption is high—not low. In contrast, glycine acts as a "trigger lock" on macrophages, preventing their activation to the M1, proinflammatory functionality. Teasing out cause versus effect in the measurement of these nutrients and metabolites is not an easy task, especially when the "control" or "normal" levels of glycine are taken axiomatically to indicate healthy function. In particular, the Ren group appears to be entirely unaware of the research implicating glycine deficiency in the development of diabetes, and specifically in the function of macrophages, even though the very mechanism of glycine's action had been worked out by the Thurman group two decades prior.

Another amino acid that has been studied for its beneficial effects on diseases rooted in inflammation is the sulfur-containing amino acid, cysteine. Cysteine is one of the three amino acids that make up the body's main antioxidant molecule, glutathione (GSH). GSH has had a substantial following among life scientists as a key player in age-related diseases—including diabetes—and aging itself. It is attractive in that oxidative damage causes inflammation, and many believe this implicates a lack of GSH as a cause of chronic inflammation. However, oxidative damage is also ***caused by*** inflammation. For example, one of the poisons produced and secreted by activated (M1) macrophages to kill invading bacteria is hydrogen peroxide. But such poisons also kill normal cells, and cell death causes more inflammation (if glycine is deficient), and the vicious cycle becomes chronic. A team of researchers at Baylor College in Houston, Texas led by Premranjan Kumar and Rajagopal Sekhar has propounded a theory called the "power of three." By this they mean that equally important in controlling oxidative damage and chronic inflammation are GSH, and the two GSH component amino acids glycine and cysteine. Following this line of thinking they have conducted successful trials in laboratory rodents and humans using supplementation with "GlyNAc," the combination glycine and N-acetylcysteine (a readily absorbable form of cysteine).

This line of research is puzzling, however, in that the authors make no attempt to separate the effects of these two amino acids individually; they only test the combination. This is problematic in light of our old friend Ockham's Razor, which would dictate that if the benefits of GlyNAc could be achieved by the supplementation of either amino acid alone, then the other can be dismissed as the principal player. In fact, the benefits the Kumar and Sekhar group obtain have been observed with just glycine supplementation alone. Add to that the fact that cysteine (and methionine, from which it is derived) is abundant in muscle meat, so the idea that the typical omnivorous could produce a cysteine deficiency can be dismissed. There is more to the story of GSH, however, and we'll take a deeper look at GSH and the Kumar and Sekhar team's work in a later chapter. Getting back to the case of diabetes specifically, we may again recall that the 2008 clinical study of Cruz et al demonstrated that glycine supplementation alone could reverse type 2 diabetes.

But unfortunately, most of the effort in terms of looking for substances that help to normalize islet function is in looking for drugs, especially anti-inflammatory drugs. So there has been research on NSAIDS (non-steroidal anti-inflammatory drugs), such as the aspirin-like drug salsalate, or TNFalpha inhibitor "biologics"—which are bioengineered antibodies to TNFalpha (e.g., etanercept, adalimumab) or IL-1β (e.g., canakinumab). These drugs have now become ubiquitous features of television advertising for inflammation-based conditions such as arthritis and Crohn's disease. They are necessarily injectable (because they are proteins, which would be destroyed by the digestive system if ingested orally) and increase the risk of all sorts of infections. That's because they are antibodies designed to impede the normal function of macrophages in destroying infectious microbes. How healthy can the population be if the use of drugs which weaken the immune system are widely and chronically used?

And let's not forget that all this research in recent years is in the shadow of the Mexico City clinical trial in 2008 (Cruz, et al), a successful trial of supplementation with ordinary glycine.

c) Nonalcoholic Fatty Liver Disease (NAFLD)

While we are on the subject of type 2 diabetes, it is a good point to address the very closely related co-morbidity of NAFLD. Actually, NAFLD encompasses several closely related co-morbidities, the most common being nonalcoholic steatohepatitis (NASH). These conditions are characterized by fatty accumulations, inflammation and scarring (fibrosis) in the liver, which can lead to cirrhosis and liver failure. It is also characterized by insulin resistance. NAFLD is very common, estimated to affect 25% of the world's population. The advent of metabolomics, which enables the measurement of up to hundreds of metabolites in the blood simultaneously, has provided a great deal of information about the metabolic defects associated with NAFLD. Thus, in 2018, Melania Gaggini at the National Research Council in Pisa, Italy and colleagues at the University of Turin, detailed differences in the profile of metabolites—especially amino acids—between normal and both obese and nonobese NAFLD patients. Unsurprisingly, while several amino acids were elevated, glycine in particular was lower in the patients compared to healthy controls, especially in the NAFLD patients who were also obese. As usual, it was not possible to pinpoint the causal chain of events in the development and progression of disease. In 2020, however, Oren Rom and colleagues at the University of Michigan zeroed in on glycine. Noting that "Lower circulating glycine is consistently reported in patients with NAFLD," they aimed to investigate "the causes for reduced glycine, its role as a causative factor, and its therapeutic potential." Using human liver biopsies and laboratory mice, they tested the activities of genes (using transcriptomics, which like all the other "-omics" methods, enables the measurement of many mRNAs simultaneously, to determine which genes are actively coding for the synthesis of particular enzymes) in NAFLD mice and people, compared to normal. As the Tan group in Singapore reported in patients with obesity in 2022, the Rom group also found that the synthesis of glycine was impaired in NAFLD. In particular, they identified a defect in the synthesis of glycine by one particular pathway in the liver, which utilizes the enzyme called alanine-glyoxylate aminotransferase (AGXT). There actually are many people (especially of northern European descent) with genetic variants of the gene for this enzyme. I suspect that such people, with decreased capacity to synthesize glycine by this route, may be more prone to have a dietary glycine deficiency show up as NAFLD. The Rom group took the next logical step of using glycine therapy to treat the illness in their experimental mouse model, based on experiments done back in 2004 by Mohammed el Hafidi and colleagues in Mexico City. Just as glycine feeding had protected el Hafidi's rats from developing NAFLD-like abnormalities, glycine-based therapy normalized all the abnormal biochemical indicators in the Rom group's study. Their approach was a bit different than just using glycine in that they sought out compounds—essentially small peptides (short chains of a few amino acids) to do even better than plain glycine. They settled on a compound they called DT-109, a tripeptide (three amino acid groups) consisting of two glycine units and one leucine unit. They found that this compound was even better than glycine alone in helping to regulate both sugar and fat in NAFLD mice. Of course, DT-109 is digested to release the free amino acids (mainly glycine), so it is really another way to supplement with glycine. Why Rom et al focused on synthetic compounds, however, rather than just ordinary, natural glycine, is another reflection of what I have referred to as a drug mentality in the medical and medical research communities. It should be noted that although the Rom group did not receive any funding from pharmaceutical companies, the drug orientation of researchers is very prevalent these days.

d) Micro Injuries causing chronic pain as long-term consequences of significant injuries:

It has long been known that substantial injuries, such as from war wounds or motor vehicle accidents, result in chronic pain years or decades after the initial injuries have healed. The treatment of such chronic pain is a major source of opioid addiction, as opioids are typically prescribed to treat such chronic pain. The fact that so many such people have had these pains substantially reduced or eliminated entirely via glycine supplementation has led me to embrace the following hypothesis:

Substantial injury often heals with the creation of scar tissue. Scar tissue is much less flexible than the soft tissue it replaces. Therefore, though long healed, formerly wounded joints, for example, when flexed, give rise to small tears: micro injuries that provoke inflammation and hence pain, as long as glycine is deficient.

Here are a few stories from people whose chronic pain was alleviated or eliminated by daily consumption of Sweetamine®, the glycine supplement I formulated and currently market:

"Having worked in construction, I had been suffering terribly with back and shoulder injuries for 17 years. All in all, I have had to endure six shoulder surgeries, five spinal surgeries and another dozen outpatient procedures. This left me with intense chronic pain, requiring me to take eight doses of opioids (morphine and Percocet) every day to get through the day, for the last several years. Then my brother heard you on the radio, and he said 'You oughtta try this stuff.' So I did, and after about a week of Sweetamine (two stick packs per day), I started to feel incredible relief from the pain and immobility, and have cut my use of painkillers literally in half. I feel like I have my life back." Anthony, Pico Rivera, CA

"My wife and I have been adding Sweetamine to our morning hot oatmeal for about 7 days, it gives sweetness to the dull tasting oatmeal. I had an old right ankle injury a few years back and developed chronic pain when I make a wrong step, the pain comes and goes, and I have been told by an Orthopedic

Surgeon that soon I will need arthroscopic surgery. Now I noticed that the ankle acts more stable and I feel no pain when walking." Tad, Mission Viejo, CA

I hear many such stories from Sweetamine consumers, including those who have had joint replacement surgeries:

"My wife and I have been using sweetamine for over a year now and we love it. We're 73 years old and are virtually pain free. I had shoulder replacement surgery last year shortly after starting on sweetamine. Even my surgeon was impressed with the speed of my recovery." John, Georgetown, TX

And those who were told by doctors that they would need joint replacement surgery, until they tried Sweetamine, which ended their pain and immobility:

Had three years of swelling and issues with my right knee. Some days, I couldn't walk at all, or had to use a walker. I finally got an MRI in May, and I was told that I needed a full knee replacement. I started taking

Sweetamine, and within two weeks there was no swelling and no more pain! Dee, Surprise, AZ

"I can't say enough good things about Sweetamine! I can move now, I don't have any pain in my knees. They told me I had osteoarthritis; that it was bone-on-bone and I would probably need knee replacement surgery, but I feel great now" Pilar, Stockton, CA

Chronic brain injury resulting from head trauma:

These days there is increasing concern about sports injuries and other forms of head trauma, as they often result in concussions and chronic encephalopathies of various sorts. Just as chronic inflammation resulting from normal cell death in brain development may produce brain abnormalities characteristic of autism, the inappropriate inflammation following traumatic injury to the brain may well be responsible for the rising tide of encephalopathies. Here again I will rely on some of my own personal self-experimentation experience to describe what does ***not*** happen, when glycine is not deficient.

Six years ago (at age 67), after allowing myself to become dehydrated, not drinking water all day and having a glass of wine after a three-hour drive, I got up quickly from a seated position and fainted. Unfortunately, I fell with my head hitting the square corner of a hardwood kitchen table on the way down. After several seconds I came to lying on the floor on my back, and in a state of some mental confusion for a minute or so. I also had a really bad headache where my head had struck the table. Fortunately, this was at the home of, and in the presence of my grown son and daughter, who quickly called 911, and I was whisked off to the hospital ER within the hour. Within two hours of the trauma, the pain at the spot of impact on my head was gone completely. The mental confusion I had upon regaining consciousness lasted no more than a minute or two. No other symptoms occurred or persisted beyond two hours at all. Nor have any occurred since.

Also, a member of my extended family had suffered a concussion from a blow to the head while playing hockey. She had recurrent headaches which persisted at least for several weeks. But after taking the Sweetamine glycine supplement for a few days, the headaches disappeared. Such chronic inflammation may well be responsible for various forms of age-related dementia, which are also on the rise.

e) "Type 3 Diabetes."

The question of brain damage resulting from chronic injury—really, from inappropriate inflammation in response to injury—brings to the fore the question of age-related dementia, such as Alzheimer's Disease. Specifically, we may ask, is age-related dementia also a result of inappropriate inflammation, due to glycine deficiency? This actually brings us back to the subject of type 2 diabetes, the association of which with dementia has recently earned Alzheimer's Disease the nickname: "Type 3 diabetes."

It is difficult to pin down the origin of the term "Type 3 Diabetes," which popped up in several places during the early years of the 21st century. Suzanne de la Monte of Brown University suggested the term in 2014; however, Zina Kroner of the American Board of Internal Medicine mentioned earlier references to the term in her 2009 review. Regardless, it is clear that many researchers have noted the endocrinological and biochemical similarities between insulin resistance and other dysfunctions observed in muscle and other tissues in Type 2 diabetes and brain abnormalities in those with Alzheimer's Disease. Moreover, the epidemiological (i.e., statistical) association between diabetes and Alzheimer's Disease has been apparent since at least 1999. This association has been the subject of several systematic reviews and meta-analyses (SRMAs), such as that published in 2009 by Feng-Ping Lu and colleagues at the National Taiwan University Hospital in Taipei. They reported a significant, overall 47% increased risk of dementia among patients with type 2 diabetes, compared to non-diabetics. Specifically, the average risk increase was 39% for Alzheimer's disease and 138% for vascular dementia. It is not surprising that both common forms of dementia correlate. (Vascular dementia is often referred to as multi-infarct dementia; a result of multiple "mini-strokes;" episodes of small blood vessel blockages in the brain. This is in contrast to Alzheimer's disease, where the primary damage is directly to the neurons of the brain themselves, rather than to their blood supply.) After all, both are characterized by (among other normalities) some form of inflammation, whether in the brain tissue itself (instigated by microglia, the brain macrophages) or macrophages in the blood and blood vessel walls.

In 2016, EM Ribe and F Lovestone of Oxford University in the UK noted: "links between insulin and IGF-1 signaling and the Alzheimer's disease-associated pathological events as well as the impact of other processes such as inflammation, oxidative stress and mitochondrial dysfunction are either common to both conditions or perhaps responsible for a mechanistic link between metabolic and neurodegenerative disease."

Now there is general agreement not only that diabetes and dementia are functionally related, but also obesity and other such "pre-diabetic" conditions, specifically insulin resistance (sometimes referred to as "metabolic syndrome") and non-alcoholic hepatic steatosis (NASH). However, there are different hypotheses proposed concerning causation, some putting more credence into aging, diet and obesity, and some others suggesting that various forms of oxidative stress may be more to blame. De la Monte, for example, makes the case that the primary insult responsible for these related conditions is the specific type of oxidative damage caused by nitrosamines. Both environmental and dietary exposure to these toxins (and known carcinogens) has increased substantially in the Western world in recent decades, along with the increased incidence of diabetes and dementia. How then, do we answer the question as to which insult is primal in the chain of events leading to these diseases?

Inflammation is increasingly recognized to be the common denominator. For example, as we have seen with the work of Cruz et al in 2008, we can definitively assign to glycine deficiency the cause of diabetes, since it can be reversed with glycine supplementation alone, and since this reversal does not involve any decrease in body weight. Rather, obesity increases the level of systemic inflammation when glycine is deficient. Hence, while obesity is clearly causally related to diabetes, it is secondary to glycine deficiency, and exacerbates glycine deficiency by reducing glycine synthesis. We have also seen that cardiovascular disease (which would also include multi-infarct or vascular dementia) is also characterized by low glycine levels. As far as oxidative damage is concerned, it is known that oxidative damage can cause inflammation, but so can virtually any kind of cellular damage provoke the activation of macrophages to cause inflammation in the presence of glycine deficiency. Ultimately, the medical/scientific community will require proof in terms of glycine supplementation's ability to prevent or reverse any or all of these inflammation-related conditions, before the glycine hypothesis is accepted.

This raises yet another question, namely, in what measure can glycine supplementation actually reverse any of these conditions? In the Cruz clinical trial, for example, it is clear that diabetes is reversed—with the average hemoglobin A1C dropping from 8.3 to 6.9 (i.e., diabetic to non-diabetic levels). However, all the patients remained on their anti-diabetic medications. Would their diabetes be completely reversed if the medications were withdrawn? This cannot be determined, of course. Yet it is reasonable to assume that what glycine supplementation achieves completely is the cessation of inappropriate inflammation. But the actual presence of disease reflects some measure of chronic damage resulting from chronic inflammation. Hence, the sooner the glycine deficiency is corrected, the less damage (especially to the pancreatic islets, where the insulin is made) and therefore, the more complete the recovery can be.

But when it comes to type three diabetes, reversal of the condition of dementia may not be possible at all, since once brain tissue is actually destroyed, it may be unable to be regenerated. Not that there is no regenerative capacity in the brain, but by the time cognitive deficits are clearly observable, the disease process has likely been going on for many years, and typically all this is happening at an advanced age, at which regenerative capacity may have been lost for the most part. Thus, in the few cases I have witnessed wherein supplementation with glycine (Sweetamine) commenced, it offered no improvement after substantial cognitive decline had been diagnosed.

4. Inflammation that is an overreaction to infection by various microbes, whether viral or bacterial

I have several examples of this, mostly from my own experience, self-observation and self-experimentation, as well as the anecdotal experiences of others. However, it all fits into the coherent picture of glycine action and inflammation that has emerged from the medical literature, as detailed above.

a) Gingivitis/Periodontitis

I had been plagued by this condition for most of my adult life. Annual or semiannual cleanings at my dentist's office were always painful and bloody affairs, with deep sulci (pockets) forming around many of my teeth, several of which were lost to infection along the way. But after daily glycine supplementation, at some point, my dental cleaning visits ceased to be painful or bloody. And the overall condition of my gums actually improved over time. From this change I conclude that the populations of bacteria that normally inhabit my mouth—and especially deep in my gums around my teeth—are simply not provoking the inflammation that characterized the "normal" state in my mouth in my pre-glycine days. Adding to the evidence of the relevance of glycine deficiency in periodontitis, in 2005 Caroline Riben-Grundstrom and colleagues in Sweden demonstrated the ability of therapeutic glycine to ameliorate periodontitis that developed after dental implants. The importance of glycine in periodontal health may also have implications for cardiovascular health, in that researchers have noted that such conditions as gingivitis are statistically linked to cardiovascular disease (CVD). This subject was reviewed in 2020 by Riccardo Nocini of the University of Verona, Italy, and colleagues. They noted that one mechanism underlying the statistical association between periodontitis and coronary heart disease (CHD) may be the translocation of bacteria to the heart and vascular tissue, and their accumulation within atherosclerotic plaques, rendering plaques more unstable and prone to cause ischemic events (heart attacks). We may hypothesize here that inflammation, which characterizes both gingivitis and atherosclerosis may be provoked unnecessarily in both locations (mouth and coronary arteries) due to glycine deficiency. This hypothesis is also supported by the Norwegian research cited earlier by Ding et al. linking adverse cardiovascular events (heart attacks) to lower glycine levels. Hence it may well be that inappropriate inflammation may be instigated by both micro-injuries in the coronary arteries as well as the presence there of otherwise harmless bacteria.

b) Shingles

It is common knowledge that shingles is an adult disease caused by an infectious agent, specifically the Herpes zoster virus, which is the same pathogen that causes chickenpox in children. Shingles is thus a resurgence of a childhood infection, the virus remaining latent for many years or decades, before symptoms and frank disease appear. The disease typically presents with a unilateral rash—typically on the lower back but sometimes occurring on the neck or face. The rash is characteristic and typically very painful, often described as the skin's being "on fire." The condition also typically lasts for a month or two, and is sometimes debilitating. There is therefore substantial promotion of shingles vaccines (e.g., "Shingrix®"), especially to older people, who have the highest prevalence.

I got a case of shingles myself in 2015, so daily eight to ten grams of supplemental glycine did not prevent shingles. And my case was classic in that it produced a characteristic unilateral rash on my lower back (with the diagnosis confirmed by my neurologist. "Classic" was the word he used to describe it.) But what was unusual was that I did not feel the rash at all. In fact I had only discovered it while bathing in the shower one morning and happened to feel the bumpy surface of the rash with my fingers. Therefore, although my shingles ran the usual course of almost two months, it never caused me any pain, although there was a mild, generalized malaise typical of many viral infections, and was therefore merely a minor nuisance. It was certainly not a disease I would ever consider using a vaccine to prevent.

This encounter with shingles led me to hypothesize that the pain usually caused by shingles is actually due to local inflammation, inappropriately set off by the infection in the absence of adequate glycine. This thinking is also in the context of hearing about similar types of relief from pain of secondary inflammation. One customer of mine suffers from periodic flare-ups of gout. Gout is a painful condition due to a metabolic imbalance that causes the crystallization of the metabolite uric acid in capillaries around joints. Hence the flare-up is mainly due to the inflammation resulting from micro-injuries caused by the uric acid crystals. But when glycine is adequate, the flare-ups produce minimal discomfort.

Another customer is a young lady who suffers from systemic lupus erythematosus (SLE). She has reported that taking daily Sweetamine enables her to flex her fingers first thing in the morning without pain. Whether glycine has any direct benefit in terms of ameliorating the disease process in SLE, which is an autoimmune disease, is unknown. My hypothesis is, again, damage caused by the disease process produces secondary inflammation when glycine is deficient.

Another such situation may be presented by allergy. I used to be exquisitely sensitive to poison ivy. Even the slightest touch of a poison ivy leaf anywhere on my skin would provoke the characteristic, delayed hypersensitivity allergic reaction: hard, white bumps that were very itchy and made me scratch them, making it all worse, of course. But sometime after taking 10g/day of supplemental glycine, it seemed that light, casual contact with poison ivy produced no major reaction.

So I tried an experiment: Handling a freshly plucked poison ivy leaf with my (gloved!) right hand, I rubbed the leaf against the back of my left hand. Afterwards I just washed my hands normally—without any special soap or solvent to wash out the poison ivy oil that persists in the skin.

Sure enough, a day or two later, the back of my left hand flared up with the typical poison ivy reaction. It looked just as ugly as any poison ivy rash had ever looked on my skin. However, the itch was minimal and actually went away soon after the rash had formed. Hence the rash never got exacerbated by scratching it. Within several days, the rash followed its typical course of disappearance, but all the while it did not itch.

c) Covid 19 Pneumonia

Back to the subject of viruses and glycine, we cannot skip over the viral pathogen of the recent pandemic, Covid 19. Covid 19 provides a perfect example of a case wherein most of the damage—life-threatening damage in millions of cases during the recent pandemic—is caused by inflammation. The severe inflammation, the pneumonia caused by Covid 19 is now typically described as a "cytokine storm" of inflammation that pervades the lungs and often kills rapidly. It is also noteworthy that, after many months of trying different treatment modalities, the standard protocol has become centered around a course of the anti-inflammatory steroid, prednisone.

From the glycine perspective, the common comorbidities are all conditions characterized by decreased glycine levels: obesity, diabetes, cancer, and most prominently, advanced age. Could it be that Covid 19 is only a life-threatening disease because it provokes inflammation in alveolar macrophages, and only when glycine is deficient?

Considering my long-standing regular consumption of supplemental glycine as an ongoing self-experiment, more results manifested when I myself became infected with Covid 19 (likely the delta strain of 2021).

As it had been with shingles, my symptoms were quite characteristic of the disease (which was confirmed by a positive rapid antigen test (Carestart). After a few days of testing positive, I had, in addition to general malaise, a fever of almost 102 F and diarrhea, both of which lasted less than 24 hours. There were also appetite changes that persisted for some weeks. The malaise persisted for two weeks and gradually subsided completely after another week. I also progressively lost my sense of smell, which was complete for a few days before gradually returning. All in all, I felt quite miserable for about three weeks, like I had a nasty intestinal virus. I needed about 12 hours per night of sleep, but was not bedridden, so overall, I would have to characterize my disease as moderate at worst.

But there was one glaring set of symptoms usually encountered with Covid that was completely absent: I had no respiratory symptoms whatsoever. No coughing, no sneezing, no wheezing, no shortness of breath. Later, in checking with various friends and family, those taking the Sweetamine glycine supplement, also had no serious disease, even those who were older (65+), and one who had even had lung surgery a year or two prior to Covid. One particular anecdote stands out: A friend of mine's dad—in his 70s—was hospitalized with Covid, and he was not doing well. His dad was on a ventilator for 12 days, as he begged the hospital not to take him off it. Finally, they discharged him and sent him home, but he was so weak, he literally could not get out of bed. I suggested that Sweetamine might help, and brought some to his house. He subsequently called me and said it was like a miracle: He fed his dad the Sweetamine in his tea in the evening, and the next morning, he just got up and went about his business!

Of course, this is only one anecdotal report. There was, however, a clinical trial of glycine administration intervention conducted by Mario Vargas and colleagues in Mexico City, on 56 severe Covid patients admitted to hospital for mechanical ventilation in the ICU, and no clinical benefit was obtained. However, the glycine was administered through a nasogastric feeding tube, and no increase in blood levels of glycine was observed. Hence it is clear that the administered glycine was not absorbed, forcing the authors to admit that a "possibility for the lack of effect of glycine was that the critical condition of patients impaired the enteral absorption of glycine and in this case, perhaps glycine administration by the intravenous route might have been a better approach."

Then of course, there is what has become known as "long covid," mainly characterized by persistent muscle weakness, but also cognitive and systemic impairments. The common denominator seems to be—you guessed it: inflammation! As of mid-2024, at least three dozen research articles have appeared in the literature about inflammation in long covid. The PHOSP-COVID Collaborative Group in the UK reported in the top medical journal *Lancet Respiratory Medicine* that only one in four patients who had been hospitalized for Covid-19 had recovered completely within five months of hospital discharge. More alarmingly, none of the patients still affected had recovered within an entire year! The authors also noted that "several inflammatory mediators were increased in individuals with the most severe physical, mental health, and cognitive impairments compared with individuals with milder ongoing impairments." They further concluded: "Both pharmacological and non-pharmacological interventions are urgently needed to improve the ongoing burden following hospitalization for COVID-19 both for individuals and healthcare systems. Our findings support the use of a precision medicine approach with potential treatable traits of systemic inflammation and obesity."

Could it be that all these patients need is supplemental glycine? The evidence certainly points in that direction, with inflammation being the central common feature of long covid. But, alas, such simple explanations seem to elude the most knowledgeable clinical experts: As Lidiane Florencio and César Fernández-de-las-Peñas of the University Rey Juan Carlos in Madrid reported almost simultaneously in the same journal (citing the same data from the PHOSP-COVID group), "Management of the post-COVID-19 condition—often referred to as long COVID—is a challenge for health-care professionals because of the heterogeneity and complexity of its clinical manifestations and the probable need for multidisciplinary management approaches." It would seem that apparent complexity immediately rules out the possibility of a simple solution for these doctors. They are likely unaware that their mindset is completely unscientific—even anti-scientific! Another iteration of this flawed logic in relation to long covid (although it is pervasive among medical science researchers and practitioners these days) was articulated in the 2023 paper by Bianca Besteher and colleagues in Jena/Magdeburg Germany. In their summary of their work on cognitive deficits in long COVID patients, they wrote: "As the heterogeneity of symptoms is increasingly recognized among long-COVID patients, it appears highly relevant to study potential pathophysiological differences among the different subtypes." After summarizing their findings in different subgroups of their patients, they conclude: "This implies a complex underlying pathomechanism in long-COVID and emphasizes the necessity to investigate the whole spectrum of post-COVID biology to determine targeted treatment strategies targeting specific sub-groups." What this thinking means, in general terms, is that when seeking to understand any set of phenomena, a multiplicity of diverse effects implies that the underlying cause is complex, and therefore the treatment must embody a diverse array of treatment strategies. But as a scientist, I look for the common factor that underlies all the diverse manifestations. They already know and acknowledge the common factor of inflammation, but even this is seen as bewilderingly complex. Just imagine if a meteorologist, in seeking to understand the weather phenomenon of precipitation, believed that one needed to separately study rain and snow and sleet and hail, as if they had nothing in common! My hypothesis, based on inflammation being found as a culprit in every study of long COVID is that Long COVID is indeed just another glycine deficiency disease (as is acute COVID pneumonia).

d) Viral infections in general

In fact, my Covid experience led me to expand my hypothesis that had been suggested by my shingles experience: Many viruses—perhaps viruses in general—are really relatively harmless unless glycine-deficiency is present. It is intriguing to consider that so many potentially deadly viruses do not universally cause serious disease, or no disease at all. Consider for example, the hepatitis B virus, and C virus, infection with which is often asymptomatic until liver cancer develops years or decades later. What about human papilloma viruses (HPV) which can cause cervical cancer years later—or not? What about HIV, long feared as a death sentence, which also does not produce disease among a significant fraction of infected individuals? Though not generally known, this was also the case for the viral disease, polio (poliomyelitis).

I was among the first generation to receive the Salk polio vaccine to prevent the scourge that was rampant until the 1950s in the United States. But in fact, before the vaccine was available, most American kids got polio. However, it was typically experienced as just another cold. Could it be that glycine was the difference between an insignificant infection and death or lifelong paralysis? Even the dreaded Ebola virus also produces no disease (only immunity) in a majority of those infected. Are viruses in general no more than a nuisance in the absence of glycine deficiency? Even more generally, a host of bodily irritations including viral and bacterial infections, and also environmental insults, such as cigarette smoking, asbestos inhalation, are only harmful because they cause injuries, which in turn, provoke inflammation. This can cause serious disease and even death when acute, such as in Covid 19 infection, or ultimately cancers when chronic.

5. Inflammatory conditions with unknown causes

Migraine headache

Headaches may follow on traumatic injury, with long-term consequences, as we have discussed. But there is also a very common type of headache not apparently related to traumatic injury, but nevertheless a product of inflammation, namely, migraines. Migraines are now known to be a specific consequence of one of the major inflammatory cytokines we have seen before in this book: Tumor

Necrosis Factor alpha (TNFα). Elevated concentrations of TNFα have been observed in the blood of patients with migraine since the 1990s, and V. Covelli and colleagues at the University of Naples, Italy, have correlated the presence of bacterial marker molecules (lipopolysaccharides, characteristic of bacterial cell walls) with TNFα in migraine as well. Hence there is the suggestion that the initial inflammation-provoking insult might be a low-grade bacterial infection. It is also particularly interesting that several inflammatory conditions are seen as comorbidities in migraine. As observed by Amrit Sudershan and colleagues in Srinagar, India, in their extensive 2024 review of TNFα in migraine: "Patients with psoriasis, inflammatory bowel diseases (IBD), and rheumatoid arthritis have a higher risk of migraines." They also detailed the partial successes with anti-TNFα drugs in recent years as well as the progress made in understanding the sequence of events in the activation of the microglia (brain macrophages) to cause inflammation.

But unfortunately, as we have seen before, Sudershan et al. seem to be swept up in the apparent complexity of the inflammatory basis of migraine, rather than seeking a common molecular denominator; a simple solution. "Migraine is a complex disorder," they say, as they highlight what they say is the need for "continued research for the discovery of genetic, epigenetic, and molecular biomarkers, which help for identifying the diseases and also motivate personalized medicine to treat migraine." Never does the word glycine enter their discussion, a fact which is curious as they defer to the classical understanding of migraine as expressed by Rami Burstein and co-workers at Harvard, including the fact that the "lower threshold of neuronal hyperexcitability, is the reason for the migraine attack initiation." I say curious, since, as I explained earlier in this book, one of glycine's natural roles as an inhibitory neurotransmitter in the brain is to prevent neuronal hyperexcitability. Then again, one should not be too surprised to see clinicians (again!) fall into the non-scientific trap of complexity, given even the title of one of Burstein et al.'s classic papers on migraine in 2015, published in the Journal of Neuroscience: "Migraine: multiple processes, complex pathophysiology."

One can find glycine appearing sporadically in the migraine literature, however, in the context of measuring CNS neurotransmitters in cerebrospinal fluid, blood or saliva. Importantly, glycine is not only an inhibitory transmitter in the CNS, but also a co-activator (co-transmitter) with glutamate, which together act as excitatory neurotransmitters. Back in the 1990s, JF Rothrock et al. at the University of California, San Diego (UCSD; 1995),.Z Alam et al. at the University of Birmingham in the UK (1998) and C Rajda et al. at the Albert Szent-Györgyi Medical University in Szeged, Hungary (1999) measured elevated levels of glycine and glutamate in CSF (cerebrospinal fluid; the fluid which circulates inside the brain and spinal cord), blood plasma and saliva, respectively in patients with migraine. Those findings of course raise the question as to which role of glycine—neuroexcitatory or neuro inhibitory—is represented by elevated levels in migraine patients. It also raises the treatment-oriented question as to whether supplementation with glycine will make things better or worse.

As to the former question, the likely answer emerged in research conducted by Martina Curto at Harvard and her colleagues in Rome, Italy. They looked at metabolites in the kynurenine pathway, a sequence of biochemical conversions of molecules of the amino acid tryptophan, which occur in the brain. In this pathway, tryptophan is converted into the amino acid kynurenine, at which point there is a fork in the road. One path produces kynurenic acid, while another results in the formation of anthranilic acid. In migraine patients, Curto and colleagues found that most of the kynurenine was converted to anthranilic acid, thus drastically reducing the formation of kynurenic acid. Importantly, kynurenic acid is a key inhibitor of glycine's action as an excitatory co-activator. Consequently, its reduction results in the increase of excitatory transmission. In other words, the natural inhibition of glycine's excitatory role is reduced in migraine. This supported the authors' hypothesis that migraine is a disorder in which excitatory neurotransmission is increased.

As to the latter question, there seems to be nothing in the literature regarding glycine supplementation in migraine, but anecdotally, my own daughter who suffered migraines a few times per year, no longer gets them at all, taking 8 grams of glycine per day. And some of my Sweetamine customers have reported dramatic benefits in their unsolicited testimonials:

Annie in Cave Creek, writes:

"THANK YOU. Love your product. I've been a chronic sufferer of migraines for most of my life. I am a 35 year old mom of three who has a very healthy diet,

exercises, and keeps my weight down, but I couldn't get my headaches to improve. I wrecked my stomach using Advil like candy to get by, but it never cured or removed my headaches. I knew my headaches were inflammation related because of stemming from my neck muscles and feeling like I always needed to pop my back. I saw numerous specialists and chiropractors but nothing helped. I changed my diet to reduce inflammation, but still, they persisted. I started taking Sweetamine and within days I felt like my headaches were gone. I thought NO WAY— too good to be true. Week THREE NO head aches. I am a believer! I am elated with this product. I tell everyone I can about it."

Dustin from Charlotte, NC writes:

"I first heard of your product on the radio. At the time I was skeptical at best about its efficacy, but I have suffered from a lot of inflammatory illness (Celiacs Disease, Leaky Gut, Spondylosis etc.), and I

thought I would try out your product for a couple of weeks to see if it would help with back and joint pain. What I did not expect was the effect it would have on my migraines. To give a little context, for the last fifteen years I have suffered from increasingly severe migraines that have been diagnosed as hemiplegic, where one side of my face and body will occasionally become paralyzed for short periods of time. At the time I ordered my first package of Sweetamine, I was on average getting one to two migraines a week, if I was lucky. Since I have been using Sweetamine every day, I don't think I have had one serious Migraine in four months. For me your product has worked wonders, completely curing my migraines, and freeing me from a great deal of back and joint pain. Thank you for creating this product. "

Note also how the benefits of glycine came as a surprise to Dustin, confirming what I mentioned earlier about the correlations between known inflammatory diseases and migraine.

b) Inflammatory Gastrointestinal disease

Inflammatory conditions such as Crohn's disease are very prevalent nowadays, but there is nothing in the recent literature I can find that references glycine in this context. One would think that glycine supplementation might help, and I have received some unsolicited testimonials from Sweetamine customers suffering from such conditions. Thus, Mackenzie of Cleveland, Ohio, writes:

"I have Crohn's and arthritis, for which I had been taking steroids for some time. But the steroids were causing so many side-effects, I went off them. But then, my arthritis got so bad that I would wake up in the morning and everything hurt really badly: my arms, my legs, everything. Then I tried Sweetamine. I took two packets a day, and within two days, most of the pain was gone. It was amazing; A-MAZING!!"

CB in Wisconsin writes:

"I have suffered from chronic diarrhea for the past ten years. By the third day taking Sweetamine I began to see a 100% improvement and now have regular bowel move-

ments. I also had a pain that ran down my right leg that has decreased considerably. I love your product!"

Mary, in Ridgefield, Washington, writes:

"I have been taking Sweetamine for almost two months. I had suffered from Irritable Bowel Syndrome (IBS) and Sweetamine calmed my intestinal tract. I also found relief from a bad shoulder that I couldn't lift above my head without pain and even had discomfort driving. Now it is almost 100% No pain!"

FM in Gilbert, Arizona, writes:

"I am thrilled with Sweetamine! I have suffered from colitis and diverticulitis for years. I have been using Sweetamine for 4 months, the pain is gone. I have regular healthy bowel movements now. The yeasty film that used to be on my tongue which I tried to get rid of for years is now gone! I have clear whites in my eyes which used to be cloudy. The arthritis in my knees is much less, too."

c) Other diseases.

Notice how all of these people who found relief from inflammatory bowel conditions also experienced relief from bodily aches and pains related to arthritis. When glycine is deficient (evidenced here by the correction of the condition with glycine supplementation), inflammation is often systemic, and shows up in all sorts of places. The last testimonial above also referenced improvement in the eyes, and the eyes are also where a host of inflammatory processes show up as disease, for example, in age-related macular degeneration (AMD).

Lucien, in Lincoln, Rhode Island, writes:

"I'm diabetic type 2 and I've been using Sweetamine for about a year, and this stuff WORKS! I have diabetic neuropathy from my knees down and severe macular degeneration (AMD). I've been taking Lyrica and Avastin for those conditions. But with Sweetamine, my doctor was able to lower the dosage of both of them and I feel a lot better. Sweetamine takes the feeling of severe tightness out of the neuropathy. It really works!"

Here again, note the presence of inflammation-related comorbidity, specifically type 2 diabetes and diabetic neuropathy. Note also the powerful prescription drugs: Lyrica (pregabalin) is a potent drug for nerve pain. According to the Cleveland Clinic, it "works by calming overactive nerves in your body." That kind of sounds like what glycine does, as one of the body's natural inhibitory neurotransmitters, doesn't it? Avastin (bevacizumab), is one of the expensive, semi-synthetic "mab" drugs (monoclonal antibodies) that need to be taken by injection. It is prescribed for AMD for the same purpose as it is prescribed for the treatment of several types of cancer, specifically, to inhibit the growth of blood vessels, thus inhibiting the growth of cancerous tumors, and also the overgrowth of blood vessels in the retina which is part of the AMD disease process.

Hence, this drug is known as an anti-angiogenic drug. And even though angiogenesis is not an inflammatory process, it is actually inhibited by glycine. Endothelial cells—the cells that line the blood vessels—are equipped with the same glycine receptors that are found in macrophages and neurons. First described in 2001 by the Thurman group at UNC, Mark McCarty of Natural Alternatives International in San Marcos, CA—a rare glycine supplementation enthusiast—suggested exploiting this fact for preventing or treating cardiovascular disease back in 2009.

Chapter Eleven
Genetic Disorders Affecting or Affected by Glycine

There are a number of rare genetic defects that affect glycine synthesis or metabolism, as well as some that cause disease without affecting glycine directly, but the severity of which may be affected by the usual state of glycine deficiency. We have already seen, from the work of Hartiala et al. at the Cleveland Clinic, that a genetic defect in the synthesis of urea (CPS-1 deficiency), which is largely made from glycine, results in higher levels of glycine and a lower incidence of heart attack in women who have this genetic defect. (This type of benefit due to having one copy of a defective gene is well known in sickle cell anemia/sickle cell trait, wherein having one copy of a defective hemoglobin synthesis gene confers resistance to malaria. Such genes are called pleiotropic genes.)

Nonketotic hyperglycinemia

Another genetic defect in glycine metabolism which is usually fatal (if the defect is complete) in the early years of life is called non-ketotic hyperglycinemia, or NKH. This involves a defect in the gene that codes for what is known as the glycine cleavage system, which is the main route of glycine metabolism. It is basically a neurological disorder, because glycine cleavage is critical within the neurons of the central nervous system, where glycine serves as a neurotransmitter. (Treatment strategies for NKH generally involve feeding with substances such as benzoic acid, which binds to glycine and helps to clear it through the urine.) In particular, the toxicity of the excess glycine is to the NMDA receptor, for which glycine acts as a co-activator (as opposed to the glycine receptor, by which glycine acts as an inhibitory neurotransmitter). Curiously, this illness does not show up until after birth. I also find it curious that the steroid DHEA (a natural product of the adrenal gland, and which is available over-the-counter as a supplement) is found to be protective of NMDA receptors in other contexts, and DHEA concentration is normally very high prenatally (as it is made then by the feto-placental unit), but drops abruptly to zero right after birth. DHEA is a precursor to the sex hormones estrogen and testosterone, but is inactive on its own. (Moreover, there are now synthetic derivatives of DHEA available, such as BNN27, which cannot be converted to sex hormones.) DHEA naturally starts to rise at about age eight (an event called adrenarche, a precursor to puberty), and it can naturally reach very high concentrations. Hence I hypothesize that in individuals with NKH, the NMDA receptor is naturally protected by DHEA before birth, such that prenatal brain development is normal. Therefore, I would suggest that supplementing with DHEA after birth might prevent the disease process of NKH from manifesting.

Since we have now veered off the subject of glycine deficiency per se, let us now extend our excursion a bit to discuss two interesting examples of heritable (genetic) diseases that have nothing at all to do with glycine, but for which glycine supplementation may offer a good treatment option. Importantly, I am not suggesting using glycine as a therapeutic drug; rather, as a nutrient which is normally deficient in the diet which can therefore be normalized by supplementation. I would even be so bold as to suggest that there are certain genetic diseases which are not really genetic diseases, but genetic variants which will only produce disease—or serious disease—in the presence of glycine deficiency.

Hereditary Hemorrhagic Telangiectasia

HHT is a genetic disorder that has long (since the mid-19th Century) been known as a hereditary bleeding disorder, similar to the various types of hemophilia. However, it is now known that the genetic defects (there are several variants) produce an oversecretion of the natural chemical mediator (like a hormone, but acting locally) called vascular endothelial growth factor or VEGF (discovered in 1989). What excess VEGF does is produce overgrowth of capillaries at various places in the body, which are then prone to bleeding. Nosebleeds are the most common symptom, but bleeding can occur anywhere in the body, such as in the intestines, and blood loss, anemia and heart failure can result from this very serious condition. It may afflict over one million people worldwide, and I just searched and found no less than 175 review papers on HHT: a very substantial literature devoted to the diagnosis and treatment of it. Most of the treatment protocols center on countering the proliferative effects of VEGF, i.e., through the use of anti-angiogenic drugs. Recall from the previous chapter that anti-angiogenic drugs are part of the medical armamentarium against cancer, for they inhibit tumor growth by inhibiting the ability of the tumor to grow a blood supply. Recall also that the cells of the endothelium—the cells which line the interior walls of the blood vessels and which constitute target tissue for VEGF—are also equipped with glycine receptors. This has been known since at least 2001, and recall that McCarty even suggested that the resulting anti-angiogenic action of glycine could be exploited for the treatment of cardiovascular disease back in 2009. Yet, for all the massive research dedicated to the treatment of HHT, not even one suggestion of the use of glycine in HHT has been proposed! When we consider the fact that most people are glycine deficient, the question that arises to me is: Would this abnormality even produce disease among people who had truly healthy levels of glycine? It would certainly do no harm to try glycine supplementation in those suffering from HHT.

Cystic fibrosis

Unlike HHT or NKH, CF is a genetic disorder which is well known to the general public, afflicting some 89,000 people worldwide. With a tremendous focus of the medical and medical research community, mean survival of CF has risen over recent years to about 53 years of age. CF arises from one of several genetic defects in the gene for a cell membrane protein called CFTR (Cystic Fibrosis Transmembrane conductance Regulator). CF is a debilitating disease, the worst manifestation being difficulty breathing due to inflammation in the airways and lungs. It has attracted my attention because CFTR is a chloride ion channel—just like the glycine receptor. Hence, it seems to me that raising glycine concentration to a truly healthy level, by glycine supplementation, might be able to augment the balancing of chloride ions via the glycine receptor, even perhaps normalizing the condition. In other words, as I suggested earlier for HHT, CF might also constitute a group of genetic variants which only produce disease—or serious disease—in the presence of a glycine deficiency. Yet, a review of CF in the prestigious Journal of the American Medical Association (JAMA) just published in 2023 makes no mention of this possibility, even though a 2017 clinical trial of glycine supplementation obtained substantial benefits in the trial patients. The 2017 randomized, placebo controlled crossover trial (Glycine-treated and placebo-treated patients switched places after eight weeks.) was conducted by the Vargas group in Mexico City, who supplemented patients (13 children aged 6-18) with 0.5 g/kg body weight/day.

The Vargas study was undertaken to test glycine as anti-inflammatory therapy, considering the known, anti-inflammatory effects of glycine, and as a bronchial muscle relaxant, since bronchial muscles are known to possess glycine receptors which induce relaxation. The subjects improved on glycine therapy, both in terms of respiratory function and the concentration of inflammatory markers in the patients' serum and sputum. The Vargas team concluded, "Thus, if replicated, our results suggest that glycine might constitute a novel therapeutic tool in patients with CF."

Appropriately, they also acknowledged that, "Larger and longer-term clinical trials, perhaps without a cross-over design, are needed to corroborate our results." Okay, that was 2017. Where are the follow-up trials for the treatment of this debilitating, life-shortening terrible disease, with a natural nutrient that is completely harmless (and inexpensive)? Such deafening silence has become all too familiar concerning treatments that are not patentable, and therefore unlikely to attract substantial investment for clinical research.

Chapter Twelve
Cutting Through TMI on Chronic Disease

Understanding glycine's role in regulating inflammation, and inflammation's role in causing all manner of chronic illness, it is easy to conclude that eating more glycine is all one needs to do to ward off or even reverse the vast majority diseases that make us sick and die these days. But then we are also confronted with voluminous research about other factors that appear to be causally important. What about the gut microbiome and the importance of probiotics? What about all the colorful plant pigments and bioflavonoids that seem to have beneficial effects? As I have pointed out repeatedly in this book, a genuinely scientific approach seeks to find the simple, central or fundamental causes of disease, rather than to focus on the complexity. But once a common, fundamental factor is identified—glycine—it also behooves us to determine and to explain how other, seemingly fundamental factors fit into the scheme we have elucidated—or not. Science tells us there is but one truth, and that everything about the issue that is true should fit into that singular scheme. So what about these other factors that are gaining traction through recent research?

The gut microbiome

More and more attention has been focused recently on the importance of the trillions of bacteria that normally inhabit the human gastrointestinal tract. Exactly what sort of mix of bacteria and what their secreted products do for us and to us is unarguably important. For many decades we have known, for example, that Vitamin K, which is essential to human life and health, is not made by the human body. (That's why it's called a vitamin.) Rather, it is made by gut bacteria, and it is an essential component of the enzymatic machinery in the liver that makes several of the clotting factors in the blood (i.e., Factors II, VII, IX and X). When we take antibiotics to combat a bacterial infection, many of the "good bacteria" in the gut are also killed, leading to a functional deficiency of Vitamin K. That's why it is not uncommon for transient bleeding issues, such as nosebleeds, to occur during or after a course of antibiotic treatment. However, there is much that is still unknown and just recently being elucidated about what these gut bacteria make and how they affect human health.

Currently there is emerging a sizable literature on the subject of the metabolite trimethylamine oxide (TMAO), which is made by gut bacteria from the nutrient choline (also from related nutrients betaine and carnitine). More precisely, gut bacteria produce the metabolite trimethylamine, which the liver oxidizes to form TMAO). Muscle meats are rich in choline so meat-eaters eat a lot of it, and have lots of gut bacteria that can utilize choline for their nutrition. At this point, readers with some knowledge of biochemistry might just be thinking: "Wait a minute! Can't choline be turned into glycine? So doesn't eating more choline mean eating more glycine?" Yes, but in the same way that eating more muscle meat also means eating more glycine, as we saw here in a previous chapter, the net result is glycine depletion because of meat's high methionine content. That's because choline is rich in methyl groups (It has three of them), and it uses up a molecule of glycine to remove each of those methyl groups. In fact, it is removal of those methyl groups that is part of the process of regenerating methionine when it is in short supply. Getting back to TMAO specifically, it has been specifically linked to the development of atherosclerosis by many researchers worldwide. How does TMAO cause cardiovascular disease (CVD)? By provoking inflammation in the blood vessel walls! Hence it is reasonable to hypothesize that the primary cause of CVD is a deficiency of glycine, for this deficiency makes the macrophages prone to initiate inflammation inappropriately.

There is also substantial literature on the effects of the increased consumption of nitrates and nitrates, inorganic substances used as preservatives in processed meats. These substances are metabolized by gut microflora to produce many nitrosamines, which are known human carcinogens. Nitrosamines are also generated by consumption of tobacco products. But how do they induce cancer? First of all, the very formation of carcinogenic nitrosamines in the body is closely tied to inflammation, via the reactive oxidation products of inflammation—Remember, that's one way inflammation kills infectious microbes. Hence we can see the operation of a vicious cycle of inappropriate inflammation, resulting in oxidative damage and oxidative formation of carcinogens, and thus more inflammation resulting from the cellular damage, that is, if glycine is deficient.

Other dietary factors and environmental exposures

Here I have mentioned only a few lines of research linking so many different types of dietary and environmental exposures to the ever-increasing prevalence of chronic diseases including CVD and cancer. There are also dietary deficiencies other than glycine—such as Vitamin D and magnesium—which keep emerging in the research literature. So how do we make sense of it all? Since I have concluded that, at least in regard to any condition involving inflammation, "All roads lead to glycine and glycine deficiency," it behooves me to trace these connections, even and especially where it does not seem obvious at first. I will take two very recent examples as cases in point, as this book is by no means intended to be an exhaustive analysis of everything linked to the incidence of chronic disease.

Pentadecanoic Acid (PDA)

A particularly exciting discovery over the last two decades concerns a specific fat molecule called pentadecanoic acid (PDA). This is actually a type of saturated fat—which of course has been a "dirty word" since the 1970s, with the belief that only the essential polyunsaturated fats such as EPA (eicosapentaenoic acid) and DHA (docosahexaenoic acid) were healthy to eat. To fill in a few basics here, most fats are molecules called triglycerides, made up of fatty acids and glycerin, a small sugar alcohol. Saturated fatty acids have no double bonds between the carbon atoms. The double bonds—in strategic places within the chain of carbon atoms—confer rigidity, and thus, confer specific stable shapes to the molecule. As fatty acids are a major component of cellular membranes—which are like highly specialized soap films—the proper function of the membranes requires a particular mix of types of fatty acids, in terms of carbon chain length and the presence of double bonds at specific places. Almost all fatty acids consist of an even number of carbon atoms, but a small percentage of them have odd numbers of carbon atoms, one of which is pentadecanoic acid (PDA), which has 15.

The study of PDA has been pioneered by veterinary epidemiologist Stephanie Venn-Watson, co-founder and CEO of Epitracker, a San Diego based natural pharmaceutical research company. Venn-Watson discovered, using metabolomics, that PDA was essential to the health and well-being of dolphins, which she was researching for the US Navy. Over the last few years, Venn-Watson has established PDA (aka C15:0) as an essential fatty acid, which, like glycine, is made in the human body, but in insufficient quantities. How does it work? The most important role of PDA is to make cell membranes in general more stable, i.e., less vulnerable to stresses such as oxidative damage. Hence, PDA deficiency leads to an increase in ferroptosis—the kind of explosive, pro-inflammatory cell death we discussed earlier, similar to the pyroptosis of macrophages as the final "nuclear detonation" of inflammation. In terms of disease manifestations, PDA deficiency syndrome—which Venn-Watson calls "Cellular Fragility Syndrome," "can increase susceptibilities to ferroptosis, dysmetabolic iron overload syndrome, type 2 diabetes, cardiovascular disease, and NAFLD." Sound familiar? This is basically the panoply of diseases rooted in chronic inflammation, which I attribute to glycine deficiency!

Furthermore, Venn-Watson also has shown "that C15:0 supplementation can reverse the described C15:0 deficiency syndrome." Just like glycine supplementation can reverse the chronic inflammation that results from glycine deficiency.

Okay, so which is it? Is glycine deficiency the real root of inflammation-related chronic disease, or is it PDA deficiency? Clearly, as both of these are now shown to be essential nutrients which are causally related to chronic inflammation and the chronic illnesses that result, which one is more important? To answer this question, we can examine the typical diet and dietary sources of both of these nutrients. Venn-Watson estimates that about one third of the US population is PDA-deficient, whereas I would estimate that glycine deficiency afflicts close to

100% of the population; basically, everyone except bone-broth enthusiasts (i.e., to the tune of about 8 oz or 250mL of bone broth daily) and those taking glycine or gelatin supplements (to the tune of several grams/day of glycine). Importantly, the best dietary sources of PDA are whole milk and full-fat dairy products, and meat, fish and poultry. But that would be the kind of diet that was typical two generations ago, before meat and dairy foods became largely low fat or fat-free, and/or low-cholesterol; in other words, before the food industry began removing most or all of the PDA from foods. And yet, the removal of saturated fats and cholesterol from food products was spurred by research showing the typical American 'meat and potatoes' diet was somehow producing large increases in the incidence of CVD and cancer, starting in the 1940s, when almost nobody in the US was PDA-deficient. More importantly, the putative mechanism by which cellular fragility from PDA deficiency leads to chronic inflammation, by causing excess cell death and therefore, inflammation, is straightforward. As we have seen, cellular death does not cause inflammation except when glycine is deficient. Remember also in our discussion of the causation of insulin resistance and diabetes, we looked at the fragility of larger and larger adipocytes (fat cells) as drivers of inflammation.

Why are these adipocytes so fragile, and therefore causes of cell death and inflammation? It may well be that this fragility is what makes obesity the main apparent cause of insulin resistance and diabetes, and that it is PDA deficiency that causes the fragility. Hence, I would not be surprised to see research popping up that shows that overweight diabetics might see some reversal of the diabetes upon supplementation with PDA. Moreover, cellular fragility of the macrophages themselves clearly makes glycine deficiency more critical, because glycine protects by virtue of allowing the influx of chloride ions to combat the deterioration of the membrane voltage, a process which will be exacerbated by PDA deficiency.

Magnesium

As I write now in 2024, one of the latest crazes in the nutritional field is magnesium, and the detrimental health effects of magnesium deficiency. Like all other minerals the body needs for normal function, magnesium is essential: you have to take it in the diet because your body cannot make any minerals. One of the main dietary sources cited is green leafy vegetables. Actually, this source is the most obvious because, in exactly the same way as iron makes blood and red meat red, it is magnesium that makes green plants green. Hence, a healthy diet with lots of green vegetables contains plenty of magnesium, but with many people not eating healthy diets, it is estimated that 15% of the US population is magnesium deficient.

In the body, magnesium—like many other metals, including copper and zinc—functions as a working part of dozens of enzymes; you might call them "nano-machines" that catalyze (facilitate) biochemical reactions. Therefore, if the body is deficient, nothing really works right. Not surprisingly, therefore, lack of magnesium is tied to all manner of chronic disease. It has also been linked to aging, because the protection of DNA appears to be one of its functions.

In terms of supplementation, most recommended are organically chelated magnesium supplements; that is, magnesium bound to an organic acid, like citric acid, because such chelates, as they are called, are more readily absorbed than a mineral form such as magnesium oxide. In fact there are now currently running competing prime time television ads for a magnesium supplement, specifically magnesium glycinate, from two different supplement manufacturers, Nature Made® and Qunol®. Wow, this stuff must be pretty good to see such big ad campaigns for one particular supplement among the many that these two companies sell! As I said earlier, magnesium is one of the latest nutritional crazes. But why, in particular, magnesium glycinate? As you might guess by its name, it is glycine that is the organic acid that is chelated to the magnesium in this supplement. (Glycine is both an organic acid and an organic base, like all the amino acids.) But in particular, in order to chelate the magnesium to glycine, it takes two molecules of glycine for each atom of magnesium, a very light metal (That's why it's more formally called "magnesium bisglycinate"). In fact each molecular complex of magnesium glycinate is actually 85% glycine! That's why it takes two large capsules to get 200 mg (Nature Made ®) or 300 mg (Qunol®) of magnesium: each daily serving actually contains 200 or 300 mg of magnesium and 1,333 mg or 2,000 mg (2 grams) of glycine! That is not enough glycine to completely correct the usual glycine deficiency, as most people need about eight grams of glycine per day. But it is enough for most glycine-deficient people to feel the difference in terms of decreased levels of inflammation.

Therefore I predict that sales of magnesium glycinate will grow very well in the next few months and years. And maybe; just maybe, some researchers studying magnesium supplementation will find much more dramatic health benefits with magnesium glycinate than with any other form of magnesium. And maybe the world of medical science will come to realize that the real magic nutrient is glycine after all.

Hence, the real bottom line here is that when it comes to chronic inflammation and the chronic diseases that result, many insults, from the oxidative damage caused by such chemicals as TMAO, to the cellular fragility caused by PDA deficiency, actually do play causal roles. But the very central role of glycine as the primary regulator of inflammation always seems to emerge when one examines the causal chain of events. There are many supplement products on the market—as well as both prescription and over-the-counter drugs—which claim to "fight inflammation," and they do. But glycine does not fight inflammation, it just stops inappropriate inflammation by keeping the master switch from switching on inappropriately. And as glycine is the way the body does it naturally, there are no drug side effects, because the natural function of inflammation in the face of a real infectious challenge is not compromised.

Chapter Thirteen
The Question of Biological Aging

Not surprisingly, since the inflammation-related chronic illnesses are associated with advancing age, we need to ask the question: To what extent are these conditions the result of simply aging? This question cannot be answered, however, unless we are able to define biological aging, and that's not so easy. In fact it has yet to be fully answered, even though the field of aging research has been around for decades, with several prominent journals dedicated to this issue.

There are basically two schools of thought with respect to aging that are prominent these days. One postulates that aging—as evidenced by the gradual deterioration of cellular functions, especially in the mitochondria (the organelles which function as cellular power plants)—is the result of the accumulation of products of oxidative damage. The oxidation of biological fuels, such as sugar and fat, produces toxic by-products, collectively referred to as reactive oxygen species (ROS). The chemistry here is very simple: Oxygen is itself reactive, producing substantial amounts of energy when it reacts with biological fuels. But in the process of these biological oxidations, small amounts of super-reactive forms of oxygen—called free radicals—are also produced. These ROS are so reactive that they cause damage to the normal cellular machinery. It is widely hypothesized that these damaged molecules gradually accumulate to the point where they interfere with normal cellular function, thus producing chronic disease associated with age.

The human body, however, is designed to chemically neutralize these ROS, using substances collectively referred to as antioxidants. One such is the enzyme superoxide dismutase. Some small antioxidant molecules are nutrients, such as Vitamin C, Vitamin E and a host of "bioflavonoids." But importantly, the body itself makes antioxidants, the one considered the most important being glutathione (GSH). There is a substantial school of thought that implicates low levels of GSH with the aging process. Chemically, glutathione is what is called a tripeptide, that is, it is composed of three amino acids. Proteins, of course, are large molecules made of many amino acid molecules; not just three. Glutathione is also different from proteins in that it is not made by the same process, but the three amino acids are linked together by the same type of bond as are the amino acids of proteins, and it can be broken down into its component amino acids by the same enzymes that break down proteins in the digestive system. The amino acid which is responsible for the antioxidant function of GSH is the sulfur-containing amino acid called cysteine. But importantly, one of the three amino acids in GSH is glycine. That means that GSH contains about 25% glycine by weight; about the same percentage as collagen. Hence it is a very rich source of glycine. Therefore it is worth examining whether benefits attributed to supplementation with GSH may actually be due to the glycine content of GSH.

A recent (2011-2022) series of studies on both mice and people by a group of doctors at Baylor Medical College in Houston, TX provides a good case in point. Focusing on the importance of GSH in aging and disease, P Kumar and RV Sekhar and colleagues tested old v young mice as well as old (in their 70's) v young men and women and diabetic v non-diabetic men and women for blood GSH and a host of clinical parameters, i.e., oxidative stress, mitochondrial dysfunction, inflammation, insulin resistance (aka "pre-diabetes"), endothelial dysfunction (endothelium being the cellular layer that lines the inner surface of blood vessels), genotoxicity, muscle strength, and cognition. The experimental groups were all supplemented with both cysteine (as N-acetylcysteine) and glycine. Hence the supplementation protocol was termed GlyNAC. These studies followed up on earlier studies on mice, wherein they had observed negative effects on these functions when GSH synthesis was suppressed. Importantly, the authors' rationale was based entirely on glycine and NAC serving as precursors to GSH. The GlyNAC-supplemented diet normalized all these clinical parameters. Not surprisingly, the authors attributed these beneficial results to the normalization of GSH levels, which they observed. Specifically, they concluded: "The observation that improving glutathione synthesis and concentrations resulted in a reduction in oxidant damage suggests that the primary reason for oxidative stress in aging is glutathione deficiency." However, by 2021 the authors acknowledged: "However, correction of GSH alone is insufficient to explain the magnitude and extent of improvements, suggesting that there may be other factors in play." Here, they introduced what they call "The power of three," meaning that GlyNAC supplementation actually accomplished three things: 1) correction of GSH status 2) addition of glycine, which has its own biochemical and other benefits and 3) addition of cysteine, which has its own important biochemical functions. Importantly, the authors do not mention the role of glycine in regulating macrophage function, i.e., controlling the primary step in inflammation.

Essentially, what we have here are two competing hypotheses. The Kumar-Sekhar team from Baylor looks at oxidative damage as the primary mover in age-related and inflammation-related decline of bodily functions. This makes sense in that it is well known that reactive oxygen species (ROS, the causative chemical species that cause oxidative damage) provoke inflammation. Restoring the synthesis and maintenance of adequate concentrations of the body's main antioxidant, glutathione (GSH) by supplementing with its main precursors, glycine and cysteine (as NAC), restores healthy function. But not quite, so, as noted above, they need to supplement their hypothesis with their ill-defined "power of three" argument.

In contrast my hypothesis views the anti-inflammatory function of glycine as primal. When a healthy glycine level is attained, inappropriate inflammation is shut off, thereby stopping the oxidative stress and damage (an intended consequence of inflammation, as oxidation is a major way of killing invading microbes), which also decreases the need for GSH. In other words, it is entirely possible that the beneficial clinical results obtained by the Baylor group were attributable to glycine alone. Unfortunately, I can find no evidence that the Baylor group experimented with supplementation of glycine alone, which would have answered the question. In addition, the modern dietary trend of eating more muscle meats and less bone and connective tissue exacerbates a glycine deficiency, but not a cysteine deficiency, as muscle meats are relatively rich in sulfur-containing (cysteine and methionine) amino acids. So why would cysteine deficiency be a growing trend?

In addition, in their own research, Sekhar et al saw GlyNAC supplementation increase blood cysteine by 55%, but glycine by 142%, blood glycine being much more deficient at the start (24% v 55% of controls, respectively). Moreover, the Sekhar group does not even mention the possible confounding effect of acetaminophen (e.g., Tylenol®) use, which might well be higher among older v younger individuals. That's because acetaminophen is well known to chemically destroy GSH by destroying its cysteine component.

But just as glycine alone might explain all the benefits the Baylor team observed with GlyNAC supplementation, glycine's function in regulating inflammation does not explain all the benefits of glycine. This was evidenced by a 2016 study by the Hayashi group in Japan, which compared mitochondrial function in cells derived from elderly donors v fetal tissue. They observed defects in cellular respiration in the mitochondria in the elderly-derived cells, but no difference in reactive oxygen species. In other words, the respiratory dysfunction was not due to oxidative damage as the prevailing theory of cellular aging would have it. Nor was the difference due to any accumulated damage to mitochondrial DNA. Rather, as the authors observed when doing a genomic analysis of the cells, i.e., an analysis of what genes were being expressed, they found that the respiratory defects of the elderly-derived cells corresponded to a downregulation (i.e., shutting off, a form of what is known as epigenetic regulation) of the two major genes that represent the two most important pathways for the synthesis of glycine. The authors then took the obvious step to test whether the downregulation of glycine synthesis was responsible for the respiratory deficit, by adding glycine to the cell culture medium, and Presto! Mitochondrial respiration was restored to the level of the young-derived fibroblasts! So here we observe another direct function of glycine in restoring youthful function, but having nothing to do with macrophages or inflammation or the glycine receptor which does not exist in these cells. So the bottom line is that glycine alone can explain the difference between elderly and youthful bodily function, but in more ways than we yet know!

In fact, there is another major function of glycine in relation to cellular injury and disease, independent of its action in regulating inflammation via the glycine receptor, that has been recognized for almost 4 decades. Getting back to research on Glutathione (GSH), it was known not only to guard against reactive oxygen species, but also to prevent cellular injury, necrosis and death. This is a big problem when it comes to organ transplants. When a kidney, for example, is removed from a donor for transplantation, the microscopic tubules of the kidney that filter the blood are very sensitive to damage from a lack of oxygen. The cells that form the part of the kidney nephron (nephrons being the multicellular working units of the kidney) called the proximal tubules, usually enjoy being bathed in oxygen-rich blood. Hence, they lack the biochemical machinery to generate energy and stay alive and functioning even temporarily without oxygen. However, it was found that adding GSH protects the tubules from necrosis, not by preventing inflammation, but by some unknown mechanism.

Harking back to 1987 (and to Chapter Three of this book), the Weinberg group at the University of Michigan ran experiments with GSH and its three component amino acids, glycine, glutamate and cysteine. The proximal tubular cells of their laboratory model system (using rabbit kidneys) were protected from hypoxic injury, but to their surprise, the cells were also completely protected by adding glycine alone. Importantly, neither glutamate nor cysteine alone were protective. Furthermore, they observed that the GSH was hydrolyzed (broken down) rapidly upon addition, into the three component amino acids, suggesting that intact GSH was not the active protective substance. Finally, they observed that oxidized GSH, in which the active antioxidant part of the GSH molecule, cysteine, was inactivated, was just as protective as active GSH. In other words, Weinberg et al had demonstrated that it is glycine, and glycine alone that protects against cell damage and death!

These findings essentially nullify the hypothesis of Kumar and Sekhar, which would require the presence of both glycine and cysteine (and in fact, intact GSH) to protect cells from oxidative damage and death.

But at least they are beginning to focus more on inflammation as the main driver of chronic disease.

In my own studies with the Orentreich Foundation on rats (2011) and with the National Institute on Aging (one of the NIH institutes) on mice (2019), supplemental glycine feeding was found to significantly extend lifespan.

Finally, in their comprehensive review of studies concerning the role of glycine in cell injury and death, published in 2016, Weinberg and colleagues observed: "Although glycine is considered a nonessential amino acid, its availability may, in fact, be limiting for all of its normal metabolic roles. Manipulating glycine levels in vivo to potentially elicit protective effects is feasible because circulating glycine levels in multiple species, including humans and rodents, are at or below the EC50s (0.5–0.75 mM) required for both its cytoprotective and anti-inflammatory effects. Serum glycine levels can be readily increased by either oral feeding or parenteral administration, and maximal cytoprotection is seen at 2–5 mM concentrations that are well below the 16–36 mM levels associated with acute toxicity."

Is this not really saying that the general population may be suffering from glycine deficiency, which is easily corrected by simply eating more glycine?

Inflammaging

Over the last decade or so, a new paradigm on aging has been in ascendance, one which is based on the observations of increased inflammation associated with age. The chronic low-grade inflammation has even spawned the new term "inflammaging." Indeed, literally hundreds of peer reviewed articles on inflammaging have appeared over the last several years, and the host of chronic diseases characterized by chronic inflammation and associated with age are often called ARDs (age-related diseases, e.g., type 2 diabetes, cardiovascular disease and cancer).

In addition to the fact that "inflammaging" researchers seem unaware of the key role of glycine in regulating inflammation, they typically make an arbitrary distinction between acute and chronic inflammation. They typically frame the issue of inflammation thus: "Inflammation is the body's natural first response to injury or infection. That's a good thing, but what's not good is when the inflammatory reaction becomes chronic, so it causes progressive damage to normal bodily tissues and functions." Hence, researchers dwell on the question: "What is it that makes inflammation become chronic, i.e., that prevents inflammation from naturally shutting off after its job is done?"

The real problem is that this is the wrong question! The problem is not acute v chronic inflammation, but appropriate v inappropriate inflammation. As we have seen, inflammation is never an appropriate response to sterile injury for it just does damage to normal tissues. And even when invading microbes are present, inflammation may be excessive, even to the point of causing death rapidly, as it can in Covid pneumonia.

Therefore "inflammaging" will not likely persist for long as a descriptor, in my opinion, for it conflates aging with a particular immune function, thus obscuring the actual fact of aging. After all, aging happens, even in the absence of inappropriate inflammation: Nobody lives forever. Yes, inflammation exacerbates the deterioration of the body and bodily functions that are associated with age. And this association is not just happenstance. Since the body's production of glycine slows with advancing age, chronic inflammation increases with advancing age. But in order to understand the process of aging per se, we need to distinguish it from the process of inflammation.

Chapter Fourteen
Why Glycine?

What makes glycine so fundamental to good health? Or we might ask, of all the essential molecules out there, why is it glycine, in particular, which is so important to cellular function, especially immune cell function? In a way, the answer has always been obvious. In fact, the critical importance of glycine has always been obvious to scientists investigating the origin of life itself. To them, the question was: What was the source of the organic molecules in the early earth before life arose on this planet that enabled organic living things to emerge?

Speculation as to the extraterrestrial origin of the organic molecules essential for life to exist on earth was long fueled by the discovery of amino acids—including glycine—in meteorites. However, the possibility that these organic compounds arose through chemical reactions here on earth after the meteorites crashed could never be ruled out. In particular, the identification of glycine was a sought-after prize because glycine is the simplest of the amino acids actually used to make proteins. Therefore, glycine was the key "missing link" between inorganic chemistry (anywhere in the universe as well as here on earth) and the biochemistry of living organisms. Glycine is also the most abundant amino acid in the body and participates in more key biochemical reactions than any other organic compound in the body.

But technological advances more recently have enabled the actual collection of material from the tail of a comet. This was the main objective of the Stardust Mission, launched back in 1999 by the US National Aeronautics and Space Administration (NASA). In 2004 the spacecraft conducted a "flyby" of the comet Wild-2, when it collected dust from the comet's tail. In 2006, the spacecraft returned to earth and safely landed in the Utah desert. Analysis of the collected "stardust" revealed a host of organic molecules—including glycine.

Nevertheless, chemical changes occurring during the return of the spacecraft could not be ruled out. Efforts to identify glycine spectroscopically in interstellar clouds and comet tails from earthbound detectors came up negative, however, as the chemical nature of glycine makes it very rare in the gaseous state, which is necessary for spectroscopy. At least, it was not until the Rosetta orbiter, which had been launched by the European Space Agency (ESA) in 2004, became the first spacecraft to actually orbit a comet, in this case, Comet Churyumov–Gerasimenko, in

2014. Finally, in 2014 the ROSINA spectrometer aboard the Rosetta orbiter was able to actually and unequivocally measure the existence of glycine in the comet's tail. This discovery was detailed in a 2016 paper by Kathrin Altwegg of the University of Bern, Switzerland, and a large international group of colleagues, in the prestigious journal *Science*.

In terms of fundamental building blocks of all living systems, glycine is about as fundamental as it gets. And as we have seen, recent research has revealed that glycine is "the alpha and the omega" of inflammation, that immune system process which is so dysregulated in the population as to be responsible for most of the diseases that make people sick and die these days. That is, glycine regulates both the initial activation of macrophages to start the inflammatory process, and the final cell membrane rupture of pyroptosis, when a macrophage "goes nuclear."

As for me, I still hold to the belief that the esteemed institutions of science and medicine, will eventually emerge from their current confusion of complexity and realize how the biggest biochemical secret to health and longevity turns out to be among the very simplest of biochemicals, hidden, as it were, in plain sight.

References Cited

Alam Z, et al. Plasma levels of neuroexcitatory amino acids in patients with migraine or tension headache. *J Neurol Sci* 1998;156:102-106. doi: 10.1016/s0022-510x(98)00023-9.

Alkhouri N, et al. Adipocyte apoptosis, a link between obesity, insulin resistance, and hepatic steatosis. *J Biol Chem* 2010;285:3428-3438. doi: 10.1074/jbc.M109.074252. Epub 2009 Nov 24.

Altwegg K, et al. Prebiotic chemicals-amino acid and phosphorus-in the coma of comet 67P/Churyumov-Gerasimenko. *Sci Adv* 2016 May 27;2(5):e1600285. doi: 10.1126/sciadv.1600285. eCollection 2016 May.

Besteher B, et al. Cortical thickness alterations and systemic inflammation define long-COVID patients with cognitive impairment. *Brain Behav Immun* 2024;116:175-184. doi: 10.1016/j.bbi.2023.11.028.

Brind J, et al. Induced abortion as an independent risk factor for breast cancer: a comprehensive review and meta-analysis. *J Epidemiol Community Health* 1996;50:481-496.

Brind J, et al. Induced abortion as an independent risk factor for breast cancer: a systematic review and meta-analysis of studies on south Asian women. *Issues Law Med* 2018;33:32-54.

Brind J. Re: Autism spectrum disorder: advances in diagnosis and evaluation (letter). ay 2018)Cite this as: *Br Med J* 2018;361:k1674. Available at https://www.sweetamine.com/images/autism2018.pdf.

Brind J, et al. (2011). Dietary glycine supplementation mimics lifespan extension by dietary methionine restriction in Fisher 344 rats. The FASEB Journal 2011;25:528 (abstract).

Burstein R, et al. Migraine: multiple processes, complex pathophysiology *J Neurosci* 2015;35:6619-29. doi: 10.1523/JNEUROSCI.0373-15.2015.

Cookson BT and Brennan MA. Pro-inflammatory programmed cell death. *Trends Microbiol.* 2001;9:113-114

Covelli V, et al. Are TNF-alpha and IL-1 beta relevant in the pathogenesis of migraine without aura? *Acta Neurol (Napoli)* 1991;13:205-211.

Cruz M, et al. Glycine treatment decreases proinflammatory cytokines and increases interferon-□ in patients with Type 2 diabetes. *J Endocrinol Invest* 2008;31: 694-699.

Curto M, et al. Altered kynurenine pathway metabolites in serum of chronic migraine patients. *J Headache Pain* 2015:17:47. doi: 10.1186/s10194-016-0638-5

de la Monte SM. Type 3 Diabetes is Sporadic Alzheimer's disease: Mini-Review
Eur Neuropsychopharmacol. 2014;24:1954–1960.
doi:10.1016/j.euroneuro.2014.06.008.

Ding Y, et al. Plasma Glycine and risk of acute myocardial infarction in patients with suspected stable angina pectoris. *J Am Heart Assoc*. 2016;5:e002621 DOI: 10.1161/JAHA.115.002621

do Prado WL, et al. Preliminary evidence of glycine as a biomarker ofca rdiovascular disease risk in children with obesity. *Int J Obesity* 2023;47:1023–1026; DOI 10.1038/s41366-023-01354-w

El Hafidi M, et al. Glycine intake decreases plasma free fatty acids, adipose cell size, and blood pressure in sucrose-fed rats. *Am J Physiol Regul Integr Comp Physiol* 287: R1387–R1393, 2004.
doi:10.1152/ajpregu.00159.2004.

Fathalla, MF. Incessant ovulation—a factor in ovarian neoplasia? Lancet. 1971;2:163. doi: 10.1016/s0140-6736(71)92335-x.Florencio LL and Fernández-de-Las-Peñas C. Long COVID: systemic inflammation and obesity as therapeutic targets *Lancet Respir Med* 2022;10:726-727. doi: 10.1016/S2213-2600(22)00159-X.

Friedlander AM. Macrophages are sensitive to anthrax lethal toxin through an acid-dependent process. *J Biol Chem* 1986;261:7123-7126.

Gaggini M. Altered amino acid concentrations in NAFLD: Impact of obesity and insulin resistance. *Hepatology* 2018;67:145-158. doi: 10.1002/hep.29465. Epub 2017 Nov 17.

Gan Z, et al. Glycinergic signaling in macrophages and its application in macrophage-associated diseases. *Front Immunol.* 2021;12:762564. doi: 10.3389/fimmu.2021.762564.

González-Ortiz M, et al. Effect of glycine on insulin secretion and action in healthy first-degree relatives of type 2 diabetes mellitus patients. *Horm Metab Res* 2001;33:358-360. doi: 10.1055/s-2001-15421

Guasch-Ferré M, et al. Metabolomics in Prediabetes and Diabetes: A Systematic Review and Meta-analysis. *Diabetes Care* 2016;39:833–846. DOI: 10.2337/dc15-2251.

Hartiala JA, et al. Genome-wide association study and targeted metabolomics identifies sex-specific association of CPS1 with coronary artery disease. *Nature Commun* 2016;7:10558. DOI: 10.1038/ncomms10558.

Hashizume O et al. (incl J-I Hayashi). Epigenetic regulation of the nuclear-coded GCAT and SHMT2 genes confers human age-associated mitochondrial respiration defects. *Sci Rep* 5:10434. DOI: 10.1038/srep10434.

Hotamisligil GS, et al. Adipose expression of tumor necrosis factor-alpha: direct role in obesity-linked insulin resistance. *Science*. 1993;259(5091):87-91. doi: 10.1126/science.7678183.

Jain M, et al. Metabolite profiling identifies a key role for glycine in rapid cancer cell proliferation *Science* 2012;336:1040-1044. DOI: 10.1126/science.1218595.

Kayagaki N. Inhibiting membrane rupture with NINJ1 antibodies limits tissue injury *Nature* 2023;618:1072- 1077. DOI 10.1038/s41586-023-06191-5.

Khan AP, et al. (including Arul Chinnaiyan) The role of sarcosine metabolism in prostate cancer progression. *Neoplasia* 2013;15:491-501. doi: 10.1593/neo.13314.

Kroner Z. The relationship between Alzheimer's disease and diabetes: Type 3 diabetes? *Altern Med Rev* 2009;14:373-9

Kumar P, et al. (incl. RV Sekhar) Glycine and N-acetylcysteine (GlyNAC) supplementation in

older adults improves glutathione deficiency, oxidative stress, mitochondrial dysfunction, inflammation, insulin resistance, endothelial dysfunction, genotoxicity, muscle strength, and cognition: Results of a pilot clinical trial. *Clin Transl Med* 2021;11:e372.
https://doi.org/10.1002/ctm2.372.

Kumar P, Osahon OW and Sekhar RV. GlyNAC (Glycine and N-Acetylcysteine) supplementation in mice increases length of life by correcting glutathione deficiency, oxidative stress, mitochondrial dysfunction, abnormalities in mitophagy and nutrient sensing, and genomic damage. Nutrients 2022;14:1114.
https://doi.org/10.3390/nu14051114.

Lee AS , et al. Developmental microglial priming in postmortem autism spectrum disorder temporal cortex. *Brain Behav Immun* 2017;62:193-202. doi:
10.1016/j.bbi.2017.01.019.

Loomis WP, et al. Diverse small molecules prevent macrophage lysis during pyroptosis. *Cell Death and Disease* 2019;10:326-335. doi: 10.1038/s41419-019-1559-4.

Lu F-P, et al. Diabetes and the risk of multi-system aging phenotypes: A systematic review and meta-analysis
PLoS ONE 2009;4:e4144

Luitjens J. NSAIDs may worsen arthritis inflammation.

Proc Radiol Soc NA, 2022 (abstract). Available at: https://press.rsna.org/pressrelease/2022_resources/2379/abstract.pdf

Luka Z, et al. (Incl Conrad Wagner). 5-Methyltetrahydrofolate Is Bound in Intersubunit Areas of Rat Liver Folate-binding Protein Glycine *N*-Methyltransferase. *J Biol Chem* 2007;282:4069-4075. DOI 10.1074/jbc.M610384200

MacMahon B, et al. Age at first birth and breast cancer risk. *Bull World Health Organ* 1970;43:209-221.

McCarty MF, et al. The hyperpolarizing impact of glycine on endothelial cells may be anti-atherogenic *Med Hypoth* 2009;73:263-264. doi: 10.1016/j.mehy.2008.12.021.

Miller RA et al. (Incl J Brind) Glycine supplementation extends lifespan of male and female mice *Aging Cell* 2019;18:e12953. doi: 10.1111/acel.12953.

Mori H, et al. Amelioration by glycine of brain damage in neonatal rat brain following hypoxia-ischemia. *Pediatr Int* 2017;59:321-327. doi: 10.1111/ped.13164

Ness RB and Cottreau C. Possible role of **ovarian** epithelial inflammation in **ovarian** cancer. *J Natl Cancer*

Inst 1999;91:1459-1467. doi: 10.1093/jnci/91.17.1459.

Newgard CB, et al. A branched-chain amino acid-related metabolic signature that differentiates obese and lean humans and contributes to insulin resistance. *Cell Metab* 2009;9:311–326. doi:10.1016/j.cmet.2009.02.002.

Nocini R, et al. Periodontitis, coronary heart disease and myocardial infarction: treat one, benefit all. *Blood Coagul Fibrinolysis* 2020;31:339-345. doi: 10.1097/MBC.0000000000000928.

Orentreich N, et al. Low methionine ingestion by rats extends life span. *J Nutr* 1993;123:269-74. doi: 10.1093/jn/123.2.269.

Patterson PH. Maternal infection and immune involvement in autism. *Trends Mol Med* 2011;17: 389–394. doi:10.1016/j.molmed.2011.03.001

PHOSP-COVID Collaborative Group. Clinical characteristics with inflammation profiling of long COVID and association with 1-year recovery following hospitalisation in the UK: a prospective observational study. *Lancet Respir Med* 2022;10:761-775. doi: 10.1016/S2213-2600(22)00127-8. Epub 2022 Apr 23

C Rajda, et al. Amino acids in the saliva of patients with migraine. *Headache* 1999;39:644-649. doi: 10.1046/j.1526-4610.1999.3909644.x.

Ribe EM and Lovestone S. Insulin signalling in Alzheimer's disease and diabetes: from epidemiology to molecular links. J Intern Med 2016;280:430-442. D oi: 10.1111/joim.12534.

Riben-Grundstrom C, et al. Treatment of peri-implant mucositis using a glycine powder air-polishing or ultrasonic device: a randomized clinical trial. J Clin Periodontol 2015;42:462-469.
doi: 10.1111/jcpe.12395.

Rom O, et al. Glycine-based treatment ameliorates NAFLD by modulating fatty acid oxidation, glutathione synthesis, and the gut microbiome. *Sci Transl Med.* 2020;12(572): doi:10.1126/scitranslmed.aaz2841.

Rothrock JF, et al. Cerebrospinal fluid analyses in migraine patients and controls. *Cephalalgia* 1995;15:489-493. doi: 10.1046/j.1468-2982.1995.1506489.x.

Schemmer P, et al. Glycine reduces platelet aggregation. *Amino Acids* 2013;44:925–931
DOI 10.1007/s00726-012-1422-8

Schmidt JA, et al. Plasma concentrations and intakes of amino acids in male meat-eaters, fish-eaters, vegetarians and vegans: a cross-sectional analysis in the EPIC-Oxford cohort. *Eur J Clin Nutr* 2016;70:306–312

Sreekumar A, et al. (incl. Arul Chinnaiyan) Metabolomic profiles delineate potential role for sarcosine in prostate cancer progression. *Nature* 2009;457:910-914. doi:10.1038/nature07762

Stute P, et al. Reappraising 21 years of the WHI study: Putting the findings in context for clinical practice. *Maturitas* 2023;174:8-13. doi: 10.1016/j.maturitas.2023.04.271.

Sudershan A, et al. Role of TNF-α in the Pathogenesis of Migraine Pain Res Manag 2024;2024:1377143. doi: 10.1155/2024/1377143. eCollection 2024.

Tan HC, et al. De novo glycine synthesis is reduced in adults with morbid obesity and increases following bariatric surgery. *Front. Endocrinol* 2022;13:900343. doi: 10.3389/fendo.2022.900343

Vargas MH, et al. Effect of oral glycine on the clinical, spirometric and inflammatory status in subjects with cystic fibrosis: a pilot randomized trial. *BMC Pulm Med* 2017;17:206. doi: 10.1186/s12890-017-0528-x.

Vargas MH, et al. Glycine by enteral route does not improve major clinical outcomes in severe COVID-19: a randomized clinical pilot trial Sci Rep 2024;14:11566. doi: 10.1038/s41598-024-62321-7.

Vaughan DE PAI-1 and atherothrombosis. *J Thromb Haemost* 2005;3:1879-1883. doi: 10.1111/j.1538-7836.2005.01420.x.

Venn-Watson S. The Cellular Stability Hypothesis: Evidence of Ferroptosis and Accelerated Aging-Associated Diseases as Newly Identified Nutritional Pentadecanoic Acid (C15:0) Deficiency Syndrome. *Metabolites* 2024;14:355. doi: 10.3390/metabo14070355.

Weisberg SP, et al. Obesity is associated with macrophage accumulation in adipose tissue *J Clin Invest* 2003;112:1796-1808. doi: 10.1172/JCI19246.

Wheeler MD, et al. (incl. RG Thurman). Glycine: a new anti-inflammatory immunonutrient. *Cell Mol Life Sci* 1999;56:843–856

Weinberg JH, et al. Cytoprotective effects of glycine and glutathione against hypoxic injury to renal tubules. *J Clin Invest* 1987;80:1446–1454. DOI: 10.1172/JCI113224.

Weinberg JH, et al. The role of glycine in regulated cell death. *Cell Mol Life Sci* 2016;73:2285–2308. doi:10.1007/s00018-016-2201-6.

Ying W, et al. Role of macrophages in obesity-associated islet inflammation and beta cell abnormalities. *Nat Rev Endocrinol* 2020 February ; 16:81–90. doi:10.1038/s41574-019-0286-3.

Zhao F. Inflammation: The Body's 'Fire' Burns Threats,

but Things Go Wrong When It Smolders. *Epoch Health, The Epoch Times.* August 17, 2024.

Zychlinsky A, et al. Shigella flexneri induces apoptosis in infected macrophages. *Nature* 1992;358:167–169. doi: 10.1038/358167a0

Made in the USA
Monee, IL
05 November 2026

67986810R00108